EVERYDAY THOUGHTS ABOUT NATURE

# Science & Technology Education Library

VOLUME 9



SCOPE

The book series *Science & Technology Education Library* provides a publication forum for scholarship in science and technology education. It aims to publish innovative books which are at the forefront of the field. Monographs as well as collections of papers will be published.

# Everyday Thoughts about Nature

## A Worldview Investigation of Important Concepts Students Use to Make Sense of Nature with Specific Attention of Science

*by*

WILLIAM W. COBERN

*Western Michigan University,*
*Kalamazoo,*
*Michigan, U.S.A.*

KLUWER ACADEMIC PUBLISHERS

DORDRECHT / BOSTON / LONDON

A C.I.P. Ctalogue record for this book is available from the Library of Congress.

ISBN 0-7923-6344-2 (HB)

---

Published by Kluwer Academic Publishers,
P.O. Box 17, 3300 AA Dordrecht, The Netherlands.

Sold and distributed in North, Central and South America
by Kluwer Academic Publishers,
101 Philip Drive, Norwell, MA 02061, U.S.A.

In all other countries, sold and distributed
by Kluwer Academic Publishers,
P.O. Box 322, 3300 AH Dordrecht, The Netherlands.

*Printed on acid-free paper*



Printed in the Netherlands.

***This book is dedicated with love to***
*Kathy, Alex, and Rebecca*

## Table of Contents

**Acknowledgement**

I wish to acknowledge the contributions to this work made by Ms Adrienne Gibson. Throughout all the fieldwork activities, she was a tireless and invaluable collaborator. Without her contributions to the research, I would not have been able to complete this book. I wish also to acknowledge that the research reported in this book was supported by the National Science Foundation under Grant No. RED 9055834. Any opinions, findings, and conclusions or recommendations expressed in this publication are those of the author, however, and do not necessarily reflect the position, policy, or endorsement of the funding agency.

## Foreword

Dr. Okhee Lee
School of Education, University of Miami
Coral Gables, Florida 33124

I remember my excitement and appreciation several years ago when I first read *World View Theory and Science Education Research* by Bill Cobern (1991). It was a comprehensive, theoretical discussion of worldview theory in science education. I am delighted to see that Cobern has taken the next step to provide empirical accounts of worldview in *Everyday Thoughts about Nature.*

The primary goal of the book is to understand how typical ninth grade students and their science teachers think about Nature or the natural world, and how their thoughts are related to science. In pursuing this goal, the book raises a basic question about the purpose of science education for the public:

> Should science education seek to educate "scientific thinkers" in the pattern of the science teachers? Or, should science education seek to foster sound science learning within the matrices of various cultural perspectives? (p. 3)

The answer to this question becomes clear, thanks to Cobern's excellent research and persuasive arguments. First, this research takes a humanistic approach in understanding what students and teachers think about Nature "through the language and ideas voluntarily expressed" (p. 1). The research used multi-directional prompts and encouraged students and teachers to speak freely and at length in any directions they wished. This humanistic approach is stated clearly: "[T]he research seeks to illuminate some of the various ways in which students think about Nature without judging even the most unorthodox perspectives" (p. 14).

Second, the students and teachers expressed diverse views of Nature or the natural world. This conceptual diversity is significant, considering the demographic homogeneity of students and teachers from white, middle class backgrounds. Most of the students conceptualized Nature from multiple perspectives, including aesthetic, religious, mythic, conservationist, and sometimes scientific. In contrast, the four science teachers emphasized science and what one can know about Nature through science.

Finally, considering students' diverse perspectives compared to the modernist view of science in terms of mechanicism and reductionism, Cobern speculates that this discrepancy might be a main reason for the isolation and alienation of students from science. This discrepancy is not merely a matter of conceptual distance between science and students. Fundamentally, science fails to relate to students as persons.

Throughout the book, Cobern presents convincing evidence for the answer to the question he raised in the beginning – the purpose of science education for the public is “to foster sound science learning within the matrices of various cultural perspectives” (p. 106). Good scholarship, moreover, raises as many questions as the answers it provides. I would like to address several questions for further consideration. One involves that most students volunteered little school knowledge of science. Considering context-specificity of student responses, it would be informative to ask the students to relate science with Nature at the end of the interviews. It would also be informative to select a natural phenomenon and ask the students to relate Nature with science using this example.[1] These questions could tap into students’ conceptualizations of Nature and science in multiple contexts, some more related to their everyday lives and others more related to school science.

Another question involves students’ awareness of the environment. Considering that the students are from a semi-rural, recreational area and many actively engage in outdoor activities, their concern and caring for the environment seem understandable. This suggests an activist approach to science education. Students’ interest and efforts to restore and conserve Nature can be an anchor to bring science into their everyday lives and thinking. The results of the research offer valuable insights about how to consider students’ perspectives in science curriculum and instruction.

Still another question involves the “tentative” nature of the results due to a small sample size involving 16 students and 4 teachers. The small sample size enabled Cobern to make key assertions and identify major patterns. However, a larger sample size is needed to be more confident about similarities and differences among the students, between the students and teachers, and among the teachers.

Finally, one can imagine the magnitude of conceptual diversity among students from a range of ethnic, socioeconomic, gender, and geographic backgrounds. In addition to diversity, there may be commonalties among students from different backgrounds. The results in this research offer a glimpse of such diversity.

Cobern’s two books, *World View Theory and Science Education Research* and *Everyday Thoughts about Nature,* provide complementary accounts of theoretical and empirical foundations for worldview theory in science education. While many scholars have benefited from his work, many more will continue to benefit from this book.

---

[1] I have attempted something like this in my study, "Science knowledge, world views, and information sources in social and cultural contexts: Making sense after a natural disaster," *American Educational Research Journal, 36*(2), 187-219.

# Introduction

"The proper study of mankind is man."
Alexander Pope (18th century)

"You cannot study people. You can only get to know them."
C. S. Lewis (20th century)

Alexander Pope expressed the Enlightenment ideal of broadening the Scientific Revolution to include the study of human beings not only as physical organisms but *psychological* ones as well. The scientific study of the human being flourished and eventually spawned many new and more specific disciplines. Among these one counts the scientific study of science learning and teaching. All of this is part and parcel of modernism. Without commenting on the successes and failures of modernism, suffice it to say that in many disciplines today many scholars look to very different methods for addressing the questions they have about people and their behavior. There has come an attitude shift nicely summarized in C. S. Lewis' two brief sentences quoted above. It is an attitude most clearly seen to date in feminist scholarship of which Carol Gilligan's (1982) *In A Different Voice* and the Belenky, Clinchy, Goldberger & Tarule (1986) study, *Women's Ways of Knowing: The Development of Self, Voice, and Mind* are seminal examples. If we may paraphrase and adapt from these scholars, there are voices of people that need to be heard if scholars intend to have a valid understanding of people and their behavior. The feminist scholars were of course seeking ways of making women's voices heard but the importance of their work exceeds gender issues. It is important for restoring the image of people as *persons* rather than as objects of research. As I have undertaken it, the foundational perspective of worldview research is that one must hear from students and science teachers about themselves. If we as teachers can come to a better understanding of how people – including ourselves make sense of the world (especially the natural, physical world) we should be better equipped for the task of structuring effective science learning environments. Of equal importance, we should be better equipped to monitor our own activities and our curriculum for the chauvinistic tendencies of ideology and dogmatism (Griffith & Benson, 1994).

The purpose of this book, thus, is twofold. It first presents an interpretive methodology for exploring worldview presuppositions about the natural world through the language and ideas voluntarily expressed by people – in the case of this study it is ninth graders and their science teachers. A worldview refers to the culturally dependent, *implicit*, fundamental organization of the mind. This implicit organization is composed of presuppositions that predispose one to feel, think, and act in predictable patterns. The book's approach to worldview is via Kearney's (1984) logico-structural model that posits worldview as a composite of seven fundamental and universally found categories: Self, NonSelf (or the Other), Classification, Relationship, Causality, Time, and Space. It is important to note that worldview theory is not used here for identifying and cataloging different worldviews. Worldview theory is a heuristic device for suggesting interesting questions; and the logico-structural model of worldview is a useful tool for examining the fundamental understanding of the world by which people live. Specifically, this book addresses methodology for examining the sub-category "Nature" or the natural world in the primary category of NonSelf. The methodology addresses the broad question, what is it that people think about Nature or the natural world? This question is of interest because Nature is the domain of the natural sciences. Thus, one wishes to know the characteristics of how science teachers and their students understand Nature. What concepts have scope and power in their thinking about Nature? Where does science fit into their thoughts about Nature? How is science interpreted when it has become an integral part of a person's thinking about Nature? How do science teacher and student conceptualizations of Nature compare?

The second purpose for this book is to report on the answers to these type of questions, which were developed for a fairly typical group of American high school students and their fairly typical science teachers. The students in this study are typical middle-class American, ninth grade students. The concept of "typical", however, does not mean strict objective representation, based on random sampling and a calculated N-value, used as a basis for specific generalizations as is common to quantitative research. There is no suggestion here of probabilistic frequencies and no intent to generalize beyond the type of generalizing argued for by Cronbach (1975) – a subject I take up in Chapter 3. The use of "typical" in this research is a common-sense use. These sixteen students were chosen because none of them was noticeably brilliant or noticeably dull. None of the students had noticeable personal advantages (e.g., great wealth) or disadvantages (e.g., great poverty). Are there American students not represented by this group? Obviously there are, in fact, many. But that only means that this group does not represent all American ninth graders, it does not make this group *a*typical. This group is typical in that any of the sixteen students of this study can be found on almost any high school campus in America today. The same can be said about their teachers. Hence, the ideas that these people have are worth attending to.

The book develops a broad understanding of these American high school science students as persons by describing the personal thoughts, or everyday thinking, about a question relevant to science, what is Nature? The students' thoughts are then compared and contrasted with that of their science teachers. The purpose is to gain an understanding of students' fundamental beliefs about the world on the basis that developing scientific literacy can only be successful to the extent that science finds a niche in the cognitive and cultural milieu of students. The research findings are developed as a set of assertions about Nature, science, the environment, aesthetics, religion, and gender. Readers are encouraged to check these assertions against students and teachers of their own acquaintances. What one will find, I suspect, is what this research has found; there can be much diversity amongst apparent similarity.

The book is structured with three major divisions.[2] Part I is composed of three chapters that present worldview theory in the context of science education research (Ch. 1), the concept of Nature and scientific literacy (Ch. 2), and a detailed description of the research methodology (Ch. 3). Part II begins with the summary statement of the research assertions (Ch. 4). The balance of the division is composed of four chapters that discuss in detail the assertions directly pertaining to science. Chapter 5 addresses what was said about science in the general context of Nature. Though Chapter 5 does not explicitly make this point, gender was not found to be a factor in the issues examined. Chapter 6, however, turns to those issues where gender influences were found operative. Chapter 7 examines the student ideas about the environment with specific attention to ideas grounded in religion. I have treated Chapter 7 somewhat differently from the other chapters. From the start of the project, the focus of the research was science in the context of conceptualizations of Nature. During the interviews, it became apparent that the research strategy was providing a rich set of observations on student and teacher attitudes toward the environment. Hence, I have concluded Chapter 7 with an extended comment on the research implications for the findings of this study and environmentalism.

Part II concludes with Chapter 8, which compares and contrasts how the ninth graders' conceptualized Nature with the conceptualizations of their science teachers. Part III is composed of three chapters beginning with limitations and research implications of the study (Ch. 9). Chapter 10 provides a discussion that draws together the various ideas of the project; and Chapter 11 offers a philosophical conclusion that addresses the question of where is the experience of Nature in school science. Appendices at the end of the book include the complete student and teacher narratives from the study, and the basic interview protocols used in the research.

---

[2] By permission of the publishers, portions of this book were drawn from previously published work:

Cobern, W. W. (1991). *World view theory and science education research, NARST Monograph No. 3.* Manhattan, KS: National Association for Research in Science Teaching.

Cobern, W. W. (1996). Worldview theory and conceptual change in science education. *Science Education, 80*(5), 579-610.

Cobern, W. W., Gibson, A. T., & Underwood, S. A. (1999). Everyday thoughts about nature: An interpretive study of 16 ninth graders' conceptualizations of nature. *Journal of Research in Science Teaching, 36*(5), 541-564.

The work reported in this book was conducted under the auspices of the Scientific Literacy and Cultural Studies Project (SLCSP), a project that seeks to improve science education at all levels by developing a better understanding of the role culture plays in the teaching and learning of science. The research assumes that scientific literacy is the critical purpose of science education at the school level and that literacy should be understood as the degree to which science is integrated in the everyday thinking of people. Moreover, since all instruction and curriculum are communicated in some cultural idiom the improvement of science education requires a better knowledge of those cultural idioms. SLCSP was originally funded by a grant from the National Science Foundation (RED # 9055834). The research and findings, however, are the sole responsibility of the project director and associates. For further information, see the SLCSP Home Page at <http://www.wmich.edu/slcsp/slcsp.htm>.

As the reader begins this book, please note that I am telling a story. It is the story of students who know many things only one of which is science. The students use those many things that they know to explain the world of their experiences, and amongst the students, these explanations have quite different configurations. You will see that science clearly holds no position of great privilege for most of the students, but neither is science unimportant for most of them. On the other hand – and perhaps of little surprise – science dominates the conversation of their science teachers. This raises an interesting question about the purpose of science education for the lay public. Should science education seek to educate "scientific thinkers" in the pattern of the teachers? Is that what scientific literacy means? Or, should science education seek to foster sound science learning within the matrices of various cultural perspectives?

# Part I: The Theoretical Framework

# *Chapter 1*
## Worldview Theory and Science Education Research

People are purposive, intentional beings. People are creatures of habit and yet full of surprises. People can be quite unpredictable. For these reasons and many others, it is difficult to come to know people in the sense of having a causal understanding of human behavior, which was the modernist project in education. At least this cannot be done as scientists do with moving objects such as particle or projectile motion, for example, or even with the behavior of non-human animal species. What a person can do that an object cannot is to tell you about him or herself, thus helping you to get to know this person. This is of course a different kind of knowing and it suggests that getting to know a broad range of people provides an educator with exemplars of what people in general are like. "Interpretive researchers," noted Cobern (1993a, p. 936), "do not expect that the procedures of experimental natural science can ever be used to produce general laws of education. Rather, one must come to a greater understanding of what meaning is and how it is created. Similarly, the classroom environment is not to be composed of causal variables which the teacher manipulates to foster learning, but an environment mutually shaped to fit the members of the classroom, both teacher and students." My research takes it thus as axiomatic that the more educators know about students as people the better educators will be able to teach people as students in their classrooms. Among others, Fenstermacher (1979), Hawkins and Pea (1987), Lythcott (1991), and Shymansky and Kyle (1992) have espoused similar views.

One knows from fields as diverse as theology, cognitive anthropology, and philosophy that a person's thinking is based on a set of first principles, so to speak. This is a worldview according to Cobern (1991) and it is "not merely a philosophical by-product of each culture, like a shadow, but the very skeleton of concrete cognitive assumptions on which the flesh of customary behavior is hung" (Wallace, 1970, p. 143). These assumptions, or more accurately presuppositions, exert a broad influence over one's thinking.[3] One also knows from philosophers, such as John Dewey (1976) and Nel Noddings (1993), that all experience for a person is continuous. Yet, as children grow, and certainly as adults, many learn to box off portions of their thought lives so that, for example, scientific and aesthetic know-

[3] The affect, however, at any one point is likely to be low, see Jones (1972).

ledge become separately and exclusively boxed. Science educators are well aware of the phenomenon of boxing science as *school knowledge*. This boxing or compartmentalizing phenomenon is well described as parallel collateral learning by Jegede (1997, 1998), and Aikenhead and Jegede (1999).

Described in terms of border crossing (Aikenhead, 1996) , collateral learning takes place when students find it difficult (for whatever reasons) to cross the border from their own cultural backgrounds into the culture of science. Instead, according to Aikenhead & Jegede (1999, p. 276), such students "construct scientific concepts side-by-side, and with minimal interference and interaction, with their indigenous concepts (related to the same physical event)." In their studies, Waldrip and Taylor (1999) observed the failure of students to cross from their own culture into the culture of science. Waldrip & Taylor "obtained disturbingly little evidence of the positive influence of the school view of science on young people's traditional worldviews" (p. 301).[4] Students at times, however, have good reason for this type of behavior. As noted by Lowe (1995, p. 665), "to compartmentalize the world into domains, each with an interpretive framework, is not diversity but an effective survival technique." To survive academically, at least, students must yield cognitive space for school knowledge, but should any student be coerced into the cognitive defense of compartmentalizing knowledge due to irrelevance, or a perceived personal or cultural threat? Referring to this defense as cognitive apartheid (Cobern & Aikenhead, 1998) suggests no. Cognitive apartheid as a cognitive defense works against the long term best interests of the student and of the disciplines involved. It is a fundamental premise of this book, that sound education involves the integration of ideas.

Hence, if one takes seriously the concept of worldview and the assertion that all personal experience is continuous, then one can state with considerable assurance that the beliefs and experiences students bring to the classroom influence their learning experiences in the classroom. But, it is not at all clear that teachers recognize connections amongst ideas and experiences that for any given student are quite important. Therefore, gaining knowledge of what students bring to the classroom can lead to insight on how learning environments can be more effectively designed. At this point, my research seeks to supply some of this descriptive data. Clearly, however, information could come from anywhere and, as Neil Postman (1985) is wont to say about modern culture, one could easily drown in a sea of irrelevance. To avoid this, my research is grounded in a logico-structural theory of worldview (Kearney, 1984, Cobern, 1991b) which provides direction as to what research questions to ask.

## Worldview

Religions and philosophies are often seen as providing a worldview. For example, people speak of a Christian worldview or an Islamic worldview, a constructivist

[4] The Waldrip & Taylor research was done amongst non-Western students. The reader will find beginning in Chapter 5 that this same phenomenon can be observed amongst Western students as well.

worldview, or a realist worldview. Indeed, religion can be a powerful tool used by reflective individuals for articulating a religiously informed worldview. Furthermore, religion is an especially powerful formative force on the mind of a growing child, greatly influencing the contours of a child's developing worldview. However, many environmental factors influence children and adults. Though religion influences worldview, religion itself is influenced by worldview. Consider, for example, both the significant differences and similarities between African and Western Christians, between Arabian and non-Arabian Muslims. Concerning philosophy, the distinction is clearer. Philosophy by definition is a conscious, self-reflective endeavor. The premises of philosophy are arrived at through critical thought. Quite the opposite, the assumptions of worldview are typically *implicit*, and only by the greatest effort at self-reflection does one become aware of them. For most people, worldview is not a matter of personal choice. Wallace descriptively summarized the relationship of religion and philosophy with worldview:

> a world view is not merely a philosophical by-product of each culture, like a shadow, but the very skeleton of concrete cognitive assumptions on which the flesh of customary behavior is hung. World view, accordingly, may be expressed, more or less systematically, in cosmology, philosophy, ethics, religious ritual, scientific belief, and so on, but it is implicit in almost every act. (1970, p.143)

Or to paraphrase Hiebert (1976), religion and philosophy are visible expressions of a worldview.[5]

Worldview, as used in anthropology, refers to the culturally dependent, *implicit*, fundamental organization of the mind. This implicit organization is composed of presuppositions or assumptions that predispose one to feel, think, and act in predictable patterns. Kearney refers to worldview as:

> culturally organized macrothought: those dynamically inter-related basic assumptions of a people that determine much of their behavior and decision making, as well as organizing much of their body of symbolic creations...and ethnophilosophy in general. (1984, p.1)

Worldview under girds rationality. To be rational means to think and act with reason, or in other words to have an explanation or justification for thought and action. Such explanations and justifications ultimately rest upon one's presuppositions about the world. In other words, a worldview inclines one to a particular way of thinking. According to Kearney a worldview "consists of basic assumptions and images that provide a more or less coherent, though not necessarily accurate, way of thinking about the world" (1984, p.41). A worldview defines the self. It sets the boundaries of who and what *I am*. It also defines everything that is *not me*, including my relationships to the human and non-human environments. It shapes

[5] It is an important feature of *good* education that it helps students bring cultural presuppositions to a conscious level and builds a better understanding of what are one's bedrock beliefs about the world.

one's view of the universe, one's conception of time and of space. It influences one's norms and values (Kraft, 1978, p.4).

A worldview has five functions. It explains the how and why of things, and why things continue as they do. It validates "goals, institutions, and values of a society and provides them with a means for evaluating all outside influences as well as activities and attitudes within the society" (Kraft, 1974, p. 4). A worldview reinforces people "at points of anxiety or crisis in life providing security and support for the behavior of the group" (1974, p. 5); and both encourages and prescribes behavior. A worldview is an integrator. It allows one to order and systematize sense perception. According to Kraft (1974, p. 5), "this system makes it possible for a people to conceptualize what reality should be like and to understand and interpret all that happens day by day in this framework." Finally, there is an adaptive function. A worldview is "resilient and reconciles differences between the old understandings and the new in order to maintain a state of equilibrium" (1974, p. 5). Worldview helps one maintain a sense of mental order and balance in a world of change via the dialectical interaction between our extant worldview presuppositions and environmental changes.

Hence, the driving force behind the development of a worldview is a person's need to relate to the outside world. As aptly stated by Ross (1962, p. *x*), a person's "experience is useless unless interpreted." Beginning in childhood, each person interacts with his or her physical and social environment, and through this myriad of environmental interactions, worldview presuppositions are unconsciously constructed. The process occurs over a long period of time, with the formative, childhood years being of most importance. Through the years of schooling, formal education contributes to worldview development; and in turn, a worldview provides a foundation upon which cognitive frameworks are built during the learning process. Ordinary experiences of maturation indicate that at some point of maturity (e.g., as an adult) the malleability of a worldview begins to decrease. It becomes resilient in the face of change providing an adult with cognitive stability. As noted above, however, worldviews are adaptable so as to allow even adults to adjust to new environments. Thus, while worldview presuppositions are strongly held, they are not immutable. The strength with which a mature worldview is held appears to be inversely related to the degree of heterogeneity in a culture. The more heterogeneity, the less strongly a worldview is apt to be held. This proposed process of worldview development and change is what Kearney calls "dialectical constructionism" (1984, p. 3) . It has a compelling ring to it because it shares much with Piaget's genetic epistemology (1971) as well as with constructivist learning theory (Ausubel, Novak & Hanesian, 1978; Gunstone, 1988; Tobin, 1993). In human mental architecture, worldview is the foundation upon which a person constructs cognitive and perceptual frameworks.

Cultural anthropologists study worldviews to learn more about people and their cultures. They want to know why one group acts and thinks this way, while another group acts and thinks a different way. For educators the importance of worldview is identified in two assumptions:

> the best immediate understanding of behavior is offered by understanding the thoughts that underlie the behavior, and... other things being equal, the economy of human thought and the nature of culture are such that cognitive assumptions at work in one area of life, say economic production, will also organize thinking in others, say... ideas about human nature. (Kearney, 1984, p. 3, 4)

In other words, one assumes that thought has a great influence on action; and furthermore, that even very different areas of thought are influenced by what might be called generic, cognitive presuppositions. In a discussion of conceptual change, Carey (1986, p. 1129, her emphasis) wrote, "we must find better ways of *representing* conceptual structures so as to be able to analyze conceptual reorganization." Knowing more about student worldviews should help researchers come to a better understanding of conceptual change by providing a more complete understanding of conceptual structure. It should as well enable educators to better understand student attitudes and achievement in general.[6]

Worldview research[7] in science education dates at least to Kilbourn (1984) and Proper, Wideen, and Ivany (1988). Cobern (1991b) borrowed a logico-structural model of worldview from anthropologist Kearney (1984) in an attempt to bring greater coherence and sophistication to worldview research in science education. Kearney (1984) proposed a composite model of worldview. The composite is one of seven fundamental and universally found categories that Kearny likens to the diagnostic categories used by physicians:

> Although the doctor is confronted with a variety of patients, he can presumably describe the most significant medical facts about them in terms of... features common to all patients, e.g., blood pressure, pulse, respiration. (p. 65)

While the physician's categories are filled by measurements, the worldview categories are filled by logically consistent presuppositions about reality. The categories can be briefly described as follows.[8]

> **Self and NonSelf**: "Universe" (or cosmos) is the English language term for ultimate inclusiveness. Within the universe, an individual's primary point of reference is himself or herself, i.e., the Self. The functioning of any human society is dependent upon self-identification and culturally determined notions of the nature of Self. Every "Self" (or a person's sense of self) exists and interacts within an environment, i.e., the NonSelf (everything in the universe except the Self).

---

[6] For a dissenting view on the worldview thesis, see Dzama and Osborne (1999)

[7] For a much more complete analysis of the relationship between worldview and science education, and specific examples, see Cobern (1991b & 1996).

[8] For a detailed discussion of the seven logico-structural categories, see (Kearney, 1984, Ch. 3).

**Classification**: The Self-NonSelf differentiation is the clearest and perhaps most significant example of the Classification category. Beyond this fundamental classification are many possible sub-classifications of the NonSelf into classes such as the social, the natural, and the spiritual.

**Relationship**: It is difficult if not impossible to discuss Self and NonSelf totally independent of each other. To speak of the Self invariably involves some context, i.e., the NonSelf. To speak of Nature is to invariably invoke the terms of human relationship with Nature. It is in the interactions of Self and NonSelf that the senses of Self and NonSelf form.[9]

**Causality**: Causality in this anthropological perspective refers not to the traditional philosophical study of cause and effect or to any specific scientific sense of causality. Rather, it refers to all ideas (formal and informal) that people have about cause and effect that they use to explain the various and sundry phenomena and events that take place in any aspect of life and experience.

**Time & Space**: These two categories refer to the fundamental presuppositions that people hold about Time and Space. These may be the most difficult categories to explain given that time and space are such "taken for granted" experiences. People seldom imagine that people in other cultures conceive of Time and Space quite differently. For example, most Westerners conceive of Time as linear and are quite perplexed at the idea that Time could be thought of as circular.

These seven categories offer a useful way of examining people's understanding of the world in which they live. They say nothing about the content of this understanding, however. For example, in American culture many will address the description of Self by asking, who am *I*? In Japanese culture the description of Self is much more likely to raise the question, "who are *we*?" (Kawasaki, personal communication).

Worldview theoretical work was extended in Cobern (1993b, 1996; also see Baker, 1998; Cobern & Loving, 2000a; Lewis, 1998) and applied to empirical work in Cobern (1993a & 1997), Cobern, Gibson, and Underwood (1999), George (1999), Lassiter (1993), Lawrenz and Gray (1995), Lee (1999), Lewis (1996), Ogunniyi et al. (1995), and Slay (1999). Allen (1998), Allen and Crawley (1998), Emereole (1998), Lowe (1995 & 1997), Lynch and Jones (1995), Mattson (1997),

---

[9] I have found that people often have difficulties with this idea that it is in the interaction of Self and NonSelf that the senses of Self and NonSelf form. The difficulty is usually in the form of an objection to the distinction that the logical-structural model of worldview posits between in Self and NonSelf. As one reviewer of this book's manuscript put it, "I guess I need to be convinced that the self and non-self should be so distinct – what would the transcendentalists say?" The point that must be kept in mind is that the logical-structural categories are analytical devices for studying worldview. They are not themselves the specific content of any worldview. Hence, the reviewer's question confuses terms. The ideas that transcendentalists hold about Self and NonSelf contribute to the *content* of the Self, NonSelf, and Relationship categories. The content of a transcendentalist worldview would likely show a *blending* (Relationship category) of beliefs about the Self and NonSelf. However, no culture blends the content of Self and NonSelf to the extent that the categories are indistinguishable. The simple proof of this is that there is no language that is devoid of first person singular pronouns.

Otijele (1991), Ramorogo (1998), Shumba (1999), and Waldrip and Taylor (1999) examples of related work.

Chapter 2 specifically addresses the category and subcategory that are the focal point for this study. One should understand, however, that other scholars who have an interest in worldview theory and research have adopted different models – models other than logico-structuralism – for their research. In science education, the Kawagley et al (1998)[10] article provides one example. In environmental studies, there is a research avenue concerning "environmental worldview" and related issues that takes a non logico-structural approach to worldview[11] (see Arcury, Johnson, & Scollay, 1986; Catton & Dunlap, 1978; Dunlab & Van Liere, 1978; and Kempton, Boster, & Hartley, 1995). Similarly, a further variation on worldview theory has been adopted in cross-cultural counseling research by Ibrahim and Kahn (1987), Ibrahim (1991 & 1993), and Tervino (1996); and in Black Studies by Baldwin (1985), Baldwin and Bell (1985), Baldwin and Hopkins (1990), and Jackson and Sears (1992); and in Religious Studies by Aidala (1984) and Fetz and Reich (1989).

[10] For the background in cultural studies for this article, see Kawagley (1995). For similar work, see Witherspoon (1974 & 1975).

[11] These other approaches to worldview are either more focused on a limited area of experience (such as Arcury's work on an "environmental worldview") or involve a less analytical – hence more ambiguous – concept of worldview.

# *Chapter 2*
## Nature and Science Literacy

The focus of this book is the subcategory Nature in the category NonSelf. As stated above, the NonSelf is everything in the universe except the Self. The NonSelf can be divided into domains of equivalent, nonequivalent, or hierarchical taxonomic status (see Cobern, 1991b, Ch. 4). The simplest division is into domains of human environment and physical environment, or society and Nature. Most cultures, including Western culture, fall along the lines of Redfield's (1952) tripartite division: Humanity (society), Nature, and God (the transcendent). Some of the bitterest controversies in public education can be traced to differences in the Self-NonSelf axis and domains of the NonSelf. For example, a group of citizens may believe that the education establishment is promulgating a worldview solely based on society and Nature. In opposition stands a group of educators who may believe that the citizen group is unjustly trying to promote in the schools a religious worldview. As one would expect, the aspect of the NonSelf of interest in the science classroom is Nature. From a worldview perspective, one would ask, what is the image of Nature projected in the science classroom? What is Nature like according to science instruction? There is a rich literature on what people in different societies and at different times have believed about Nature (see Glacken, 1967; Knopf, 1987).[12] Is it wise for educators to assume that students coming into the science classroom will fully accept as both appropriate and important the image of Nature projected there, when the literature indicates that there are many views of Nature? Environmentalists have taken an interest in student beliefs about Nature and perceptions of Nature (e.g., Knopf, 1987). It is an interest that should be extended to all science educators.

Sperry (1983, p. 114) suggested that Nature is,

> a tremendously complex concept that includes all the immutable and emergent forces of cosmic causation that control everything from

[12] "Nature" as a concept of has been addressed in the history of ideas by scholars such as Glacken (1967) and Simon (1970) and in cultural studies by scholars such as Nakamura (1980). The concept of Nature has been addressed from feminist perspectives. See Merchant (1989). More recently, the concept of Nature has been addressed from the perspective of environmental ethics and morality. See Nevers, Gehard, Billman-Mahecha (1997).

> high-energy subnuclear particles to galaxies, not forgetting the causal properties that govern brain function and behavior at individual, interpersonal, and social levels.

This definition has a rather reductionist flavor characteristic of modern, Western culture. The Western view of Nature is characteristically mechanistic. It is an inorganic view of the world as a "great machine, which, once it has been set in motion, by virtue of its construction performs the work for which it was called into existence" (Dijksterhuis, 1986, p. 495; also see Stillman, 1977). This mechanicism which dates to Newton posits the whole as a simple sum of its parts. Causal relations are linearly conceived and context independent. Key elements in this view are the "regularity, permanence and predictability of the universe" (Kearney, 1971, p. 24). With all due respect to quantum mechanics and the inroads made by postmodernism, mechanicism is orthodoxy and remains a pervasive view in Western culture. Foster (1935 & 1936), Glacken (1967), Lewis ([1960], 1994), Merchant (1989), Simon (1970), and Thomas (1983) are all significant contributions to the literature on Nature in Western thought.

True to their Western heritage, Americans frequently view Nature as an object for "mastery" (White, 1967)[13] – though this is beginning to change under the influence of environmental thought (Kempton, et. al., 1995). In other cultures, Nature is more likely to be valued for its beauty, if not actually held in reverence (Foster, 1991; Kawasaki, 1990 & 1996; Nakamura, 1980). These world-view differences have consequences. Watanabe noted that despite the frequency of earthquakes in Japan, it was only after contact with Westerners that the Japanese began the scientific study of earthquakes. According to Watanabe, "this can be explained largely by [the Japanese] attitude of coexisting with Nature" (1974, p. 281). American feminist literature records a similar attitude but with different effect. The Western feminist presuppositions under girding the Self-NonSelf relationship are characterized by "interrelatedness and interconnectedness, wholeness and one-ness, inseparability of observer and observed, transcendence of the either-or dichotomy, dynamic and organic processes" (Perreault, 1979, p. 4), not unlike Watanabe's description of the Japanese view of Nature. Many researchers now argue that the gap between women's ways of knowing and the traditional culture of science and science education alienates many women students (Barr & Birke, 1994). More-over, rather then interpreting this gap as a deficit among women, feminist scholar-ship such as Evelyn Fox Keller's (1983) seminal biography of Barbara McClintock, *A Feeling for the Organism*, has helped to strengthen the feminist contention that good science does not necessarily require the traditional Western view of Nature.

---

[13] White (1967) argued that the Christian theology was at the root of Western disregard for the environment. Young (1974) and more recently Harrison (1999) disputed his thesis. Moreover, the Evangelical Environmental Network, initiated by World Vision and Evangelicals for Social Action, is a movement among Christians to "respond faithfully to our biblical mandate for caring stewardship of God's creation" (see, http://www.esa-online.org/een/, and DeWitt, 1998).

The traditional Western theme of dominance raises concerns about reckless individualism and the wanton exploitation of Nature. The Indian philosopher Radhakrishnan (1967, p. 145) commented that,

> the modern mechanistic societies lack the vision of self in man. They recognize only an external mechanistic universe reflected in the machines that man has devised. This is how disintegration becomes the key image of the modern world.

In North America and Europe, especially since the 1960s, a small but growing number of people have embraced and advocated non-Western views of the relationship between Self and NonSelf (Aidala, 1984; Carpenter, 1996). Snively and Corsiglia (2000), among others, have raised the public's general understanding of the importance of traditional ecological knowledge, that is the knowledge of Nature as traditionally understood by indigenous peoples.[14] Others in North American and Europe have adopted even more radicalized non-Western views of the relationship between Self and NonSelf. Organizations such as the Animal Liberation Front and Earth First! actively seek the end not only of all animal experimentation in science, but as well an end to meat, leather, and wool industries.[15] Even more radical is Knight's (1996) *The Voluntary Human Extinction Movement*:

> VHEMT (pronounced vehement) is a movement not an organization. It's a movement advanced by people who care about life on planet Earth. We're not just a bunch of misanthropes and anti-social, Malthusian misfits, taking morbid delight whenever disaster strikes humans. Nothing could be farther from the truth. Voluntary human extinction is the humanitarian alternative to human disasters. (http://www.vhemt.org/)

These radical activists demonstrate how serious worldview differences can be. That the differences can lead to antiscientific views has not gone unnoticed among some scientists. Holton (1993), Levitt and Gross (1994), and Theocharis and Psimopoules (1987) and others have all sounded urgent warnings.

The science classroom should not be exempted from this discussion on Nature. The classroom confluence of students, teacher, and curriculum materials can easily involve a great range of worldview variation. Included in this variation one is likely to find various views on Nature, or what might be called the "nature of" Nature. Hence, from a worldview perspective one must ask what is the image of Nature projected in the science classroom? What is Nature like according to science instruction? Kilbourn (1984), Proper, Wideen, and Ivany (1988), Smolicz and Nunan (1975), Whatley (1989), Wilson (1981), and Woolnough (1989) all suggest that mechanicism is prevalent in Western science education. Is it wise for educators to assume that students will easily accept a mechanistic view of Nature as both appropriate and important when the literature indicates that there are many views

---

[14] Also see Alcoze (1991), Kidwell (1985), and Snively (1990).

[15] See Foote (1992); the *Los Angeles Times* (1989, p. A6); *The World & I* (1995, vol. 10, no. 4: 356-383).

of Nature? Indeed, the criticism of modern, Western scientific views of Nature (e.g., Merchant, 1989) provide reason to investigate the views fostered in a science class. This line of thought suggests a broad agenda for cultural studies research in science education, premised on the assertion that all ideas, including scientific ones, are expressed within a cultural setting (Geertz , 1973). Thus, one must ask how does the cultural setting of the science teacher and curriculum compare with student cultural settings? As part of that agenda, the research reported in this book addresses cognitive culture among students and their science teachers: How do they understand Nature? What concepts have scope and power in their thinking? Where does science fit into their thoughts about Nature? How is science interpreted when it has become an integral part of student thinking about Nature?

These questions, moreover, suggest an alternative view of scientific literacy and literacy assessment. The elimination of scientific illiteracy is the principal and historic objective of science education at the school level. Scientific illiteracy is typically defined as a *cognitive deficit*, to use Layton's (n.d.) and Jenkins' (1992) description, assessed by quantitative measures involving both science concepts and processes. The NAEP (1979) and Miller (1987) assessment series in the USA are good examples of this approach. Layton, Jenkins, MacGill, and Davey (1993) identified three weaknesses with this approach. The first is simply that literacy assessments involve a limited number of scientific concepts and it may well be that people taking the assessments know *other* things about science which are not on the assessment. Second, laypeople (in contrast to scientists and science educators) may have different interests and so the concepts used in the assessments are a mismatch with lay interests. Third, laypeople may have a different purpose for understanding science. The literacy assessments are based on a scientist's view of the natural world. In the public, the purpose for understanding science may have more to do with "'scientific savvy'... the practical 'street wisdom' which a citizen needs to cope effectively in an advanced industrial democracy" (Layton et. al., 1993, p. 13). With these objections in mind, the acid test of whether science has influenced the way a person thinks is not a set of explicit questions about science, such as asking for an explanation of a particular science concept or the construction of an experiment to test a scientific hypothesis. No, the acid test is whether science has become an authentic part of a person's everyday thinking.

The research reported here asks: To what extent do students enjoin scientific knowledge vis-à-vis other domains of knowledge in a discussion about Nature, given that science is unarguably relevant to the topic of Nature; and yet, Nature is a topic that most people do not explicitly associate with science? Moreover, what are the concepts that appear to have scope and force in the thinking on this topic? It is one thing to be able to give correct answers on a science exam; it is quite another to appropriately use scientific knowledge in the absence of any kind of science prompt or cue. As noted by Heller and Finley (1992, p. 259), it is "important to understand *when* and *how* students apply their knowledge" (also see Heath & McLaughlin, 1994). Thus, I intend that this research accurately represent the typical thoughts that students and their teachers in the study have about Nature, bearing in mind that as students learn and mature their ideas change and develop. The research seeks to illuminate some of the various ways in which students think about Nature without judging even the most unorthodox perspectives. My position is that scientific literacy can be developed from a number of different perspectives on Nature – only one of which is the rather narrow perspective of typical school science curricula – and for that to happen in science education there needs to be an increased appraisal of the knowledge and values brought to the science classroom from other domains.

The next chapter (Chapter 3) details the methodology used to illuminate people's "everyday" thoughts about Nature. I use the term "everyday" to indicate that what I am trying to get at are the typical, the natural thoughts of the people with whom I am conversing.

## *Chapter 3*
An Interpretive Methodology

The research objective of my methodology was to map the qualitatively different conceptualizations of Nature[16] held by people, or what might be called *terrain of belief* regarding Nature (also see Jones, 1972; Marton, 1988), and thus to better understand conceptualizations of Nature and the place science finds in those conceptualizations. The worldview methodology described here is a modified naturalistic inquiry, interview technique (Kvale, 1983; Spradley, 1979) with constant comparative analysis (Lincoln & Guba, 1990) and assertion development (Strauss, 1987). The concept of Nature, however, is quite profound and not easily addressed extemporaneously. Thus, for most persons, one cannot simply ask, "What is Nature?", and expect to learn much. One could ask a series of questions, but questions inevitably suggest certain types of answers to the exclusion of others. Instead, in this methodology, data was gathered via semi-structured interviews that involved elicitation devices (Bliss & Ogborn, 1987; Fetterman, 1989) designed to encourage a person to talk at length about Nature, to "think aloud" about Nature. So as not to lead the interview, the elicitation devices are *multi-directional* prompts; that is, each device prompts in many directions at one time. It is up to the informant to decide which of the many directions to take. At no time does an interviewer intro-duce science (or any other specific domain of knowledge) in the conversation. It was solely up to the informant to bring science or any other topic to the discussion. The interviewer does use a set of heuristic questions (discussed in the next section) that are important for uncovering scientific ideas without directly asking about science. The interviewer, consistent with Spradley (1979) and Kvale (1983), is there to ask probing questions and to encourage the informant to speak freely and at length.

[16] There are idealizations of the concept "Nature" and other uses for the word "Nature" as in "human Nature". In this study, we take Nature to mean the natural world: "The material world and its phenomena. The forces and processes that produce and control all the phenomena of the material world" (*The American Heritage Dictionary of the English Language, Third Edition*, 1992). The research method includes procedures for insuring that persons being interviewed are aware of this general definition.

**The Interview Procedure**

The actual decision to conduct qualitative interviews did not come at the inception of this research. In Japan, Ogawa (1986a & 1988) had been pursuing a similar interest in the cultural understanding of the concept of "Nature" and how such under-standings might relate to science education. Ogawa used a word association method with his Japanese students. In Japanese, the word "Nature" is translated as *Shizen* (Kawasaki, 1990 & 1996), but as Kawasaki and Ogawa both note the two words are by no means direct equivalents. Sensing the importance of how Nature (in English) and Shizen (in Japanese) are interpreted with respect to science education, Kawasaki (personal communication) asked a number of eminent science educators in various countries to comment on the meaning of the English word Nature. Whether the respondent was a native English speaker or not, all clearly recognized the relevance of Nature to science. The differences amongst the respondents lay elsewhere. Hence, my research began with a series of pilot studies patterned on Ogawa's work. American high school and college students were asked to write, in one instance, a few sentences about the meaning of Nature or the natural world. In a second instance, students were asked to write five words they associated with Nature or the natural world. I found the results interesting, to say the least, and sometimes interpretable. However, many times the results were not interpretable because the respondents had not limited their associations to Nature as Nature refers to the natural world. It was clear that many American students used the word "Nature" in ways very different from standard dictionary, philosophical and scientific definitions for this word. Ogawa (personal communication) did not encounter this problem. The situation in Japanese is that Shizen does not have the breadth of associations and usage that the English word "Nature" commonly has (at least in the USA). I mention this experience as a way to caution future researchers on the importance of attending to language/cultural factors when borrowing a research methodology.

This language usage problem was overcome by beginning an interview with a *focusing event*. The focusing event is designed to insure that an informant has a basic understanding of what the interview is about without suggesting too much about the attributes and value of Nature. Specifically, the interview begins with the informant viewing a set of six natural landscape photographs depicting Nature at micro and macroscopic levels (including outer space), and Nature as both benevolent and dangerous. People and human constructions are shown in only one photo and this is the photo intended to show the power and danger of Nature. The object and number of photographs can vary but the photographs must be carefully chosen so as not to over represent any one perspective of Nature. After given a few moments to examine the photographs, the informant is asked if these pictures are of Nature or the natural world. In my experience, the only picture that informants occasionally have doubts about is the picture of outer space. Occasionally a person is unsure that "space" is part of Nature but this is a minor point not found to interfere with the interview procedures. The informant is then asked whether the words "Nature" and "Natural world" name the same concept. Again, experience shows that most informants say they do. When an informant says they name different concepts, the researcher asks for an explanation and then decides which term is the best one to use for the rest of the interview.[17] At this point, the informant is asked the grand tour question: How would you define Nature, that is, the Natural world? The researcher may ask for clarification of things not understood, but the intention here is for the informant to give an open statement about Nature without any discussion that may inadvertently be suggestive.

After the informant's opening statement the interview proceeds with three tasks in the form of elicitation devices employed to elicit conversation beyond what the grand tour question and photographs could accomplish alone. As stated earlier, this methodology is about *hearing* from people. The concepts that one wants to hear about (Nature, in this case) are, however, quite profound and not easily addressed extemporaneously. Thus, one cannot simply ask a person, on the spot, "What is Nature?" and expect to learn much. One could ask a series of questions but questions inevitably suggest certain types of answers to the exclusion of others. Instead, the methodology uses elicitation devices, which are multi-directional prompts; that is, each device prompts in many directions at one time. It is up to the informant to decide which of the many directions to take. The elicitation devices comprise three word and sentence sets shown in Figures 1 and 3. The words used in Task 1 were drawn from the pilot studies conducted by Ogawa (personal communication) and by our group (i.e., university professor and researcher/teachers) described above. These studies provided language about Nature that actually came from students. To get a broader scope of words and sentences,[18] and to minimize any research bias, however, our research group made significant use of the literature on Nature in Western and Eastern thought (e.g., Cobern, 1991b, Glacken, 1967; Merchant, 1989; Thomas, 1983). The categories of epistemological, ontological, emotional, and status shown in both Figures 1 and 3, and the sub categories in Figure 3 are drawn from the literature and used here *only* to insure a wide variety of words and statements. As will become evident later, these categories are not used during any interview session, nor do they directly influence the analysis of the interview tapes. The elicitation device terms are interpreted on the basis of the meanings given by the informant.

Across the three tasks the device content partially overlaps allowing the informants to be persistently engaged by concepts relevant to the issues over the three tasks, thus minimizing the potential for unrecognized insincere comments. The built in overlap also allows triangular analysis of codes to improve the trustworthiness of interpretation. The idea of the elicitation devices is that an informant *thinks aloud* about Nature in response to the devices. The interviewer, consistent with Spradley (1979) and Kvale (1983), is there to ask probing questions and to

---

[17] None of the reasons given were relevant to the study, e.g., "I don't like the term 'natural world'."

[18] Particularly during the initial interview, to avoid introducing research bias it is critical that informants face a broad array of avenues or choices to freely take; hence the elicitation device is called a "multi-dimensional" prompt. A broad array greatly increases the likelihood that some aspect of an elicitation device will resonate with an informant's own ideas- ideas heretofore unknown to the researcher.

encourage the informant to speak freely and at length. With adults, the interview can be conducted in one sitting of 40 to 90 minutes. With adolescents, the interview is conducted in two sittings. Task One is done in the first sitting of 40 to 60 minutes and Tasks Two and Three are done in a second sitting of about the same length. In my experience, interviews with adolescents take longer than interviews with adults. Adolescent students tend to be less sure of their ideas and more deliberative. They simply take more time to say what they have to say.

It is crucial to note that at no time during the interview does the interviewer initiate a question or comment about science. It is solely up to the informant to bring science or any other topic into the discussion. Once an informant has spoken of science, it is of course both appropriate and necessary for the interviewer to follow up on the comment. The basic interview protocols for this research are given in Appendix C (page 159).

**Task One**

The elicitation device in Task One is a set of thirty-three words shown in Figure 1. Each word is printed on a 3X5 card. After the grand tour question, the interviewer begins by *randomly* sorting the thirty-three words into three equal groups to be shown to the informant one group at a time. This simply gives the informant a more manageable number of words to work with at one time. The interviewer spreads a group of cards on the table and asks the informant to sort the words into two groups according to which sentence, "Nature is _____" or "Nature is not _____," the informant would use a word to complete. (It helps to have these sentence starters visible as signs on the table.) If an informant wishes, a middle or undecided group is acceptable. The sorting procedure is repeated for the second and third groups of words. The interviewer then spreads before the informant all the “Nature is” words and asks the informant to form subgroups of words which represent (from the informant’s perspective) similar or related concepts with respect to Nature (see Figure 2). Up to this point, there is a minimum of interaction between the interviewer and informant. Now begins the discussion.

The informant is asked to pick a group of words (or it may be a single word) with which to start the discussion, for example, the first group in Figure 2. The interviewer simply asks what the informant would *first* like to talk about (see protocols in the Appendix C). When a group (or single word) is chosen the interviewer proceeds with questions such as: What is the thought about Nature conveyed by these words? What was the reason for forming this group of words? The interviewer asks for examples, plays "dumb" and asks for further explanations. Depending on what one sees in the word group the interviewer may pick out individual words and ask for more information (especially if words seem to conflict). Once the discussion of the first group is exhausted, the interviewer sets the group aside (but within sight) and asks the informant to choose the next group (or word). The process is repeated except now the interviewer asks why is this group second rather than first. Is there any connection between the first and second group? How are they different or alike?

| | | | |
|---|---|---|---|
| **Epistemological Description**:<br>(Reference to knowing about the natural world.) | confusing<br>mysterious | unexplainable<br>unpredictable | understandable<br>predictable<br>knowable |
| **Ontological Description**:<br>(Reference to what the Natural world is like.) | material<br>matter<br>living<br>complex<br>orderly<br>beautiful | dangerous<br>chaotic<br>diverse<br>powerful<br>changeable | unchangeable<br>holy<br>sacred<br>spiritual<br>pure |
| **Emotional Description**:<br>(Reference to how one feels about the natural world.) | peaceful | frightening<br>exciting | "just there"[19] |
| **Status Description**:<br>(Reference to what the Natural world is like now.) | "full of resources"<br>endangered | exploited<br>polluted | doomed<br>restorable |

Figure 1. Task One Terms

| **1st group** | **Later group** |
|---|---|
| chaotic<br>dangerous | understandable<br>orderly |

Figure 2. Example Word Groupings

It is important for the interviewer to remain alert for contradictions and ambiguities as the interview proceeds through the word groups. For example, in Figure 2 the right hand group may have been the seventh group discussed but it would have been important for the interviewer to call attention to the difference between the seventh group and the first. In this example, the interviewer might ask, In what sense is Nature both chaotic *and* orderly? And, while there is no specific set

[19] "Just there" and "full of resources" are colloquialisms taken from pilot interviews with students.

of questions that the interviewer asks, the following questions served as a tacit guide for questioning during an interview:

1. Can one *know* things about Nature?
2. If so, what *sorts of things* can one know about Nature and *how* do these things become known?
3. *Who* finds out these things that can be known about Nature?
4. *Why* do they (or anyone) seek to know such things about Nature?

These heuristic questions are important for uncovering scientific ideas without directly asking about science. For example, an informant might comment that, "Some people study Nature for a living." To which the interviewer would respond, "Who does this?" At this point, the informant might say that scientists do this and if so the interviewer might ask, "What do you think about science? Does science have anything to do with Nature?" As previously mentioned, however, at no time does the interviewer introduce the word "science."

Task One is complete when the above process is repeated for all words in both the "Nature is" and "Nature is not" groups (and the undecided group if used). If there are to be two sittings, the first sitting ends with the interviewer repeating the grand tour question, How would you define Nature, that is, the natural world? If there is only one sitting, then the interviewer moves on to Tasks Two and Three.

**Tasks Two and Three**

The elicitation device for both Tasks Two and Three is the same set of eighteen sentences shown in Figure 3 with each statement printed on a 3x5 card. If Task Two is done in a second sitting, the interview begins by asking the informant to recall again his or her definition of Nature. Subsequently, two signs are displayed before the informant, "Agree" and "Disagree," and the informant is shown all eighteen cards. The informant is asked to divide the cards into two groups, i.e., those with which the informant is in general agreement and those against (again, an undecided group is allowable). The informant is then asked to separately review the two groups, and from this point on the procedures are identical to those for Task One.

The sentences of Tasks Two and Three provide a significant amount of redundancy with Task One as can be seen in the categories used in both Figures 1 and 3. The sentences are more suggestive than the words in Task One. At this point, however, the informant has already established his or her preferred viewpoints through the sorting of words in Task One. The sentences of Task Two give the informant the opportunity to bring more focus to issues concerning epistemology, ontology, emotion vis-à-vis Nature, and perceptions of the status of Nature. An informant can further develop ideas because there is sustained engagement with the topics. The statement devise also provides specific prompts in some areas. This is done because some concepts cannot always be adequately represented by single words. In Task One the words holy, sacred, and spiritual are meant as religious

| | |
|---|---|
| **Epistemological Description–** Reference to knowing about the natural world. | **(a) Knowable:**<br>1. Nature is something that should be studied so that we can learn more about it.<br>2. It is important to understand how things work in Nature.<br>**(b) Unknowable:**<br>3. Nature is difficult to understand.<br>4. To me Nature is mysterious. |
| **Ontological Description–** Reference to what the Natural world is like. | **(a) Super naturalistic:**<br>5. I see in Nature the work of God.<br>6. I find in Nature a spiritual quality.<br>7. Nature is the result of purpose and things happen in Nature because of purpose.<br>**(b) Naturalistic:**<br>8. I view Nature as something solid, substantial and reliable.<br>9. Nature is the material, concrete world around us.<br>10. The Natural world is all there is, all there ever was, all there ever will be.<br>11. The material world of Nature is the only real world there is. |
| **Emotional Description–** Reference to how one feels about the natural world. | **(a) Positive:**<br>12. I see beauty in Nature.<br>13. I have a pleasant emotional response to Nature.<br>**(b) Neutral:**<br>14. Nature is an everyday part of life that I generally do not think much about. |
| **Status Description–** Reference to what the Natural world is like now. | **(a) Resource Orientation:**<br>15. Nature is a very important resource: water, energy, food, materials for making things.<br>16. Without the things that we get from Nature we could not enjoy the lifestyle we have today.<br>**(b) Conservationist Orientation:**<br>17. I believe Nature needs to be protected.<br>18. I am concerned about pollution and the damage it does to Nature. |

Figure 3. Statements for Tasks Two and Three

words but not all informants use these words in a religious way. An informant who rejected all the religious words in Task One as not applying to Nature might still pick the sentence "I see in Nature the work of God" as an important idea. The status words from Task One are treated similarly in Tasks Two and Three.

By the conclusion of Tasks One and Two, the informant has spoken at considerable length on the subject of Nature. Now the person is in a much better position to pin point the ideas about Nature that are of most importance to this person. That is what Task Three is about. Task Three is a dyad statement, ranking task. In Task Three the informant is shown random combinations of two sentences from Task Two. The informant is shown the first two randomly chosen statements and asked to discard both, keep both, or keep only one depending on how strongly the informant agrees with the statements. Then the interviewer randomly selects a third statement, which the informant compares with any statements kept from the first comparison. Again, the informant must decide whether to keep or discard the new statement. The informant may also change his or her mind about previous statements. At this point, the informant may choose to keep all showing sentences or discard one or more according to how strongly the informant agrees with the statements. This process is repeated until all 18 have been drawn. During the process, the informant is asked to keep the retained sentences in rank order. An informant has the latitude to reorder the sentences as new ones are drawn and kept. When all sentences have been drawn, the informant is asked to check the rank order one last time before the interviewer records the order number. What is left on the table are the rank ordered statements about Nature with which the informant strongly agrees. The order of rank is used to establish the order for the narratives, discussed on pages 27 to 29.

It is crucial to note that although Task Three is essentially a ranking task, the interview incorporates a think-aloud procedure. This affords the informant one more opportunity to discuss the statements and insures that the interviewer knows how the informant is interpreting the sentences. The interview ends with a repeat of the grand tour question followed by the question, "Please tell me something that you know about Nature that is quite important?"

## Generalizing

The research procedures develop what are essentially working hypotheses (not generalizations) in the form of interpretive assertions through an emergent design as advocated by Cronbach (1975), Erickson (1998), and Strauss (1987). My understanding of generalizing is much influenced by Lee J. Cronbach. Speaking to an audience of quantitative researchers, Cronbach (1975) argued that:

> Instead of making generalization the ruling consideration in our research, I suggest that we reverse our priorities. An observer collecting data in one particular situation is in a position to appraise a practice or proposition in that setting, observing effects in context. In trying to describe and account for what happened... he will give

> equally careful attention to uncontrolled conditions, to personal characteristics, and to events that occurred during the treatment and measurement. As he goes from situation to situation, his first task is to describe and interpret the effect anew in each locale... As results accumulate, a person who seeks understanding will do his best to trace how the uncontrolled factors could have caused local departures from the modal effect. That is, generalization comes late, and the exception is taken as seriously as the rule (Cronbach, 1975, p. 124-125)

It is arresting that in the midst of Cronbach's discussion he compares his suggestion for multiple-location research that attends to local differences with the concept of "thick description" from anthropology; and Cronbach actually cites Clifford Geertz (1973, Ch. 1). It is arresting, of course, because so much of qualitative research in education draws upon this same concept from anthropology. Indeed, the narratives of the present study form a thick description with the subsequent intent to develop sets of narratives over a series of different locales. Thus, this type of research is like "core sampling" in geology rather than like social science field surveys that canvass broad populations.

**Analysis Procedures**

The analysis of data begins with the transcription and coding of the interview audiotapes. Transcripts are coded by assigning code words, which represent chunks or pieces of information within a transcript, to transcript line number. Codes can be embedded within line numbers assigned to other codes, and the same set of line numbers can be co-coded with two or more codes. Some of the code words used in this research were taken from the Task One prompt words, but many came from the transcripts themselves (i.e., words used by an informant were used as codes). In addition, other codes are chosen by the researchers as needed. Whenever possible it is advisable that coding be done by a caucus of two researchers with a third researcher coding independently of the first two. Afterwards, they together analyze the coded transcript to iron out disagreements. The codes and code definitions are subsequently kept in a lexicon so that they can be used consistently throughout the coding process.

Once the transcripts are coded, a computer program (e.g., The Ethnograph or HyperQual) facilitates the sorting and printing of text segments by code words. The text printouts associated (by line numbers) with each code also list embedded codes and co-codes for the same set of line numbers. These segment printouts by code word are summarized on a worksheet (see Figure 4) that is attached to the printed segments for each code word. These sorted and collected segments allow for the review of text associated with the code words, which is important during the construction of concept maps (see Figure 5). The construction of a concept map provides an organized overview of what was said by the informant during the interview, albeit an interpreted overview. The concept map helps identify the ideas that appear to have the most importance for the informant and how various ideas are related. Major entries in the concept map (e.g., "Beautiful" in Figure 5) serve as first level entries for an outline later to be used in the narrative construction. The ideas in the concept map underneath each major idea (e.g., "Picture" in Figure 5) become the secondary entries in the outline. The importance of an idea is gauged by its appearance across all three tasks, the number of times it is mentioned in the interview, by the informant's voice inflection, and what weight it is explicitly given by the informant, and its rank in Task Three.

Using the structure provided by the concept map and using content taken directly from the transcript, the researchers write a first person interpretive narrative for each informant. The process proceeds by following the order suggested by the concept map and Task Three, and shown in the outline. Following this code order, the text under a code (e.g., Order) is copied into a text file. This is done for all the code words, following the outline, until the researcher is confident that all useable text has been incorporated and that the informant's ideas are accurately represented. Subsequently, the concept map becomes the outline on which a first person interpretive narrative is constructed. The first person interpretive narratives are composed almost exclusively from the language of the informants, with the informants' emphases, and showing the conceptual relationships specifically mentioned by the informants. Narrative construction of this type takes many iterations of cross-examining the draft narrative with the concept map, outline, and text segments. Once the process is complete, the informant is shown the concept map and narrative for review and comment. Once constructed, a concept map and narrative are shown to the informant for review and comment. After discussion with the informant, the researchers construct the final versions incorporating the informant's editing. No narrative is ever used without the informant's full affirmation of accuracy.

This circumspect method of narrative development provides a more thorough and accurate description of a person's thoughts about Nature than can be gotten by direct questioning. The notion of a "first person interpretive narrative" is unique, however, and not without controversy. Thus, I wish to emphasize two points.

> 1. Considerable effort is exerted to make sure that the narratives are as consistent with the interview data as possible. Thus, no concepts or words of substance are contributed to the narratives by the researchers. The researchers only fill in with connecting words and prepositions. The researchers assemble various portions of an interview strictly according to topics raised by the person being interviewed.
>
> 2. All narratives are "member checked" and the person interviewed always has editorial and veto power over the form and use of his or her narrative.

Final concept maps and narratives that result from this method vary in complexity and length, especially with respect to the maturity of the informant. The example in Figure 5 is relatively simple. Figure 6 shows only a partial concept map for a

| Line Numbers | Embedded in Code Words: | Searched for Code Word | Co-Codes: | Embedded Code Words: | Comments: |
|---|---|---|---|---|---|
| 23-48 | Knowable | Science | Order | Learn | |
| 56-62 | | Science | | | |
| 123-167 | Order | Science | Biology | Chaotic<br>Mysterious | |

Figure 4. Code Summary Work Sheet

physics teacher and it is clearly much more complicated. Moreover, the narrative for a science teacher or professor can run three to four pages in length. The full narratives from the present study for both students and teachers are provided in Appendices A and B.

Throughout the research process, from interviewing to coding to concept map and narrative production, the researchers keep alert for possible assertions that could describe salient features of the data. These *tentative* assertions are logged for later use. With the finalized concept maps and narratives in hand, the researcher begins the formal process of sorting, comparing, and cross checking cases (informants) by major code categories. For example, a first analysis might divide cases by gender and examine for within-group-code consistency and cross-group code differences. Other comparisons involved the examination of cases by the codes: religion, aesthetics, knowable, science, order, and conservation. This process led to a list of 37 tentative assertions in the study reported here. These tentative assertions were subsequently grouped and reduced to smaller number of semi-final assertions. The penultimate step is to cross check each semi-final assertion against each case for confirming and disconfirming data. The researcher then constructs a narrative argument for each assertion drawing upon the first person interpretive narratives of each case in the study. The final step of the analysis process is to have the assertions and arguments externally validated by one or more qualitative researchers not involved with the study.[20] These validators cross check the assertions, supporting arguments and examples against the case concept maps and narratives. The analysis ends with the researchers revising their work in light of the external validation findings, which for this study resulted in fifteen final assertions. It is at this point that the researchers discuss the implications of the research (also see Cobern, Gibson, & Underwood, 1999).

This assertion development procedure represents a change from the bipolar code analysis that was used in the first conceptualizations of Nature study (Cobern, 1993a). The assertion analysis approach is more typical of qualitative research (Gallagher, 1991; Denzin & Lincoln, 1994) and has the advantage of greater familiarity in the research community. Moreover, in recent research, the statement form of assertions was found to be more informative than bipolar codes when considering the research and instruction implications of the research.

[20] In the study reported here, three independent evaluators were used.

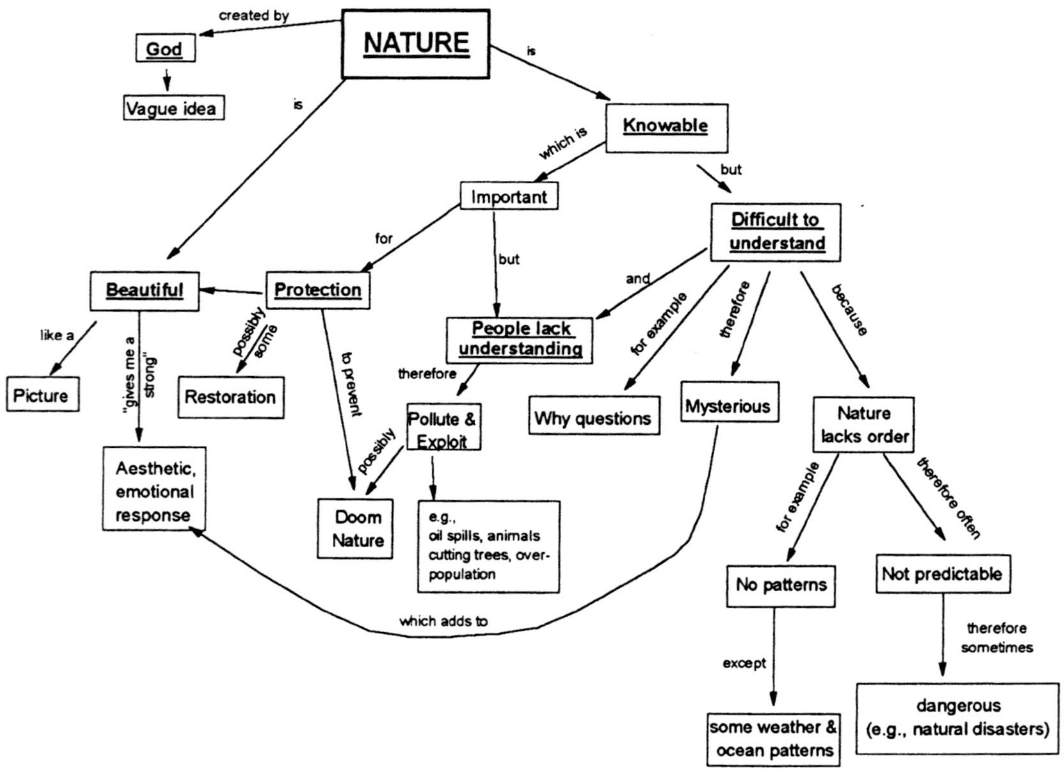


Figure 5. Example Concept Map From a Ninth Grade Student

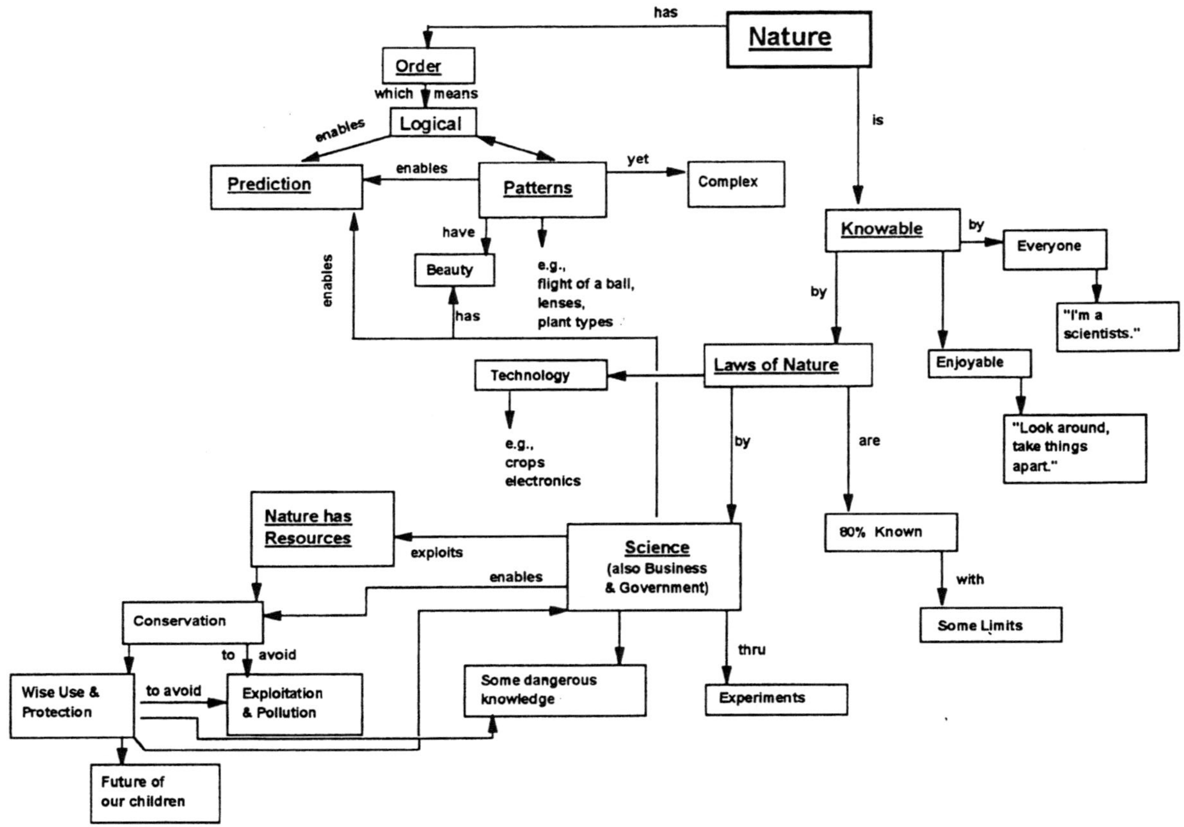


Figure 6. A High School Physics Teacher

**The Students and Their Teachers**

The research was a descriptive case study, as described by Stake (1988), of sixteen ninth graders from a suburban, semi rural high school in the central desert region of Arizona. The two science researcher/teachers (see Flake et. al., 1995; Wong, 1995) who participated in the research asked for student volunteers. From these volunteers, students were chosen so that there was a rough balance by sex and science grade success. Fourteen students were Anglo-American. Two female students were racially mixed: one with Japanese and Anglo-American ancestry and the other with American Indian and Anglo-American ancestry. All were from middle to upper middle income homes. The students who participated in the research were as a group typical of students found in this high school's ninth grade science courses (see Figure 7 for a summary description). As noted in the Introduction (p. 2), the students were also typical of many students to be found on American high school campuses.

**Suburban/rural Setting**

- Ninth graders in school & their science teachers
- Largely white population
- Secularist, Christian, nominal Christian, New Age belief systems
- English first language
- Middle and upper-middle income families
- A semi rural, recreational environment

Figure 7. Summary Description

Each researcher/teacher subsequently worked with eight students.[21] The four science teachers held science degrees in the areas they taught. Two taught physical science (one man and one women); two taught life sciences (again, one man and one woman). All four were white and attended state universities. Each of the four

[21] Our research group noted that in a previous study on the same topic fifteen cases had been sufficient to achieve code redundancy during the analysis of interview transcripts (see Cobern, 1993a.)

teachers had several or more of the sixteen ninth graders in an introductory science course they each taught. The high school is located in an area many families choose because it combines proximity to a large city with a rural desert atmosphere. It is also somewhat of an artistic community that values the natural beauty of the Arizona desert.

In the next section, Part II, the discussion of the assertions developed from this research and this group of students and teachers begins. I recommend that prior to beginning Chapter 4, the reader skip ahead to Appendices A and B. There one can read the full narratives and a brief introduction to each of the students and teachers. Reading these accounts prior to reading Part II will give the reader an enhanced understanding of the assertions and their meaning.

# Part II: The Assertions

# *Chapter 4*
## Stating the Assertions

Having read through Part I, the reader should have a good understanding of the how this research was conducted and the various purposes. At this point, it is worth repeating an earlier suggestion, and that is that the full narratives given in the appendices be read before reading the analysis chapters. Some acquaintance with the narratives will enhance the reader's participation in the analysis discussion. Here in Chapter 4, I list the fifteen assertions from the research. They are presented here so that the reader can gain an over all perspective on the issues to be discussed, prior to launching into the details of the discussions.

### Chapter 5: Science and Conceptualizations of Nature

**Assertion #1**
The ninth grade students in this study tended to discuss the natural world using concepts from religion, aesthetics, science, and conservationism. A breadth of perspectives, which only sometimes were connected, typically characterized student discussion. A few of the ninth graders demonstrated through their concept of Nature a strong interest in science and voiced specific views about the Nature of science. Visible homogeneity among the students (e.g., a classroom of mainly middle class, Anglo students) actually masked substantial variation of thought with regard to Nature, science and science related concepts.

**Assertion #2**
After nine years of schooling, the level of science integration within everyday thinking remains low among these ninth graders. In their discussions of Nature most volunteer little school knowledge of science. They are aware of school science topics such as the ozone layer, rain forests, or the Big Bang theory. Such topics are voluntarily mentioned, but usually without elaboration even when asked. The ninth graders made few references to general concepts of order or pattern in Nature. Moreover, some students showed signs of a negative perspective on science while others seemed barely to know science exists.

**Assertion #3:**
Science grade success is not correlated with the concepts these ninth graders typically choose to use in a discussion about Nature. Students with the most science grade success have not necessarily grasped fundamental concepts about Nature and science such as the concepts of order and pattern, nor do they necessarily demonstrate a scientifically informed view of Nature. Science grade success does not appear to mean that a student will understand that science is about Nature, nor does having an understanding of Nature appear to influence school science grades.

**Assertion #4**
Regardless of school science grade success, most of the ninth graders attached considerable importance to personal experiences with Nature.

## Chapter 6: Gender and Conceptualizations of Nature

**Assertion #5**
The ninth grade boys and girls of this study appeared to have few differences in the way they conceptualized Nature. Both girls and boys regarded science as an important way of thinking about Nature – actually more girls than boys. The ninth grade girls were no more likely to hold caring, environmental views of Nature than were the boys although the literature suggests that these are more typically feminine perspectives.

**Assertion #6**
Most of the ninth graders, whether boys or girls, gave some form of aesthetic response to Nature, which took one of three (sometimes overlapping) forms: a conceptual perspective form, a religious (or spiritual) emotional form, and a non-religious emotional form.

**Assertion #7**
The ninth grade girls who commented on the mysterious Nature of Nature often did so in the context of their aesthetic comments. Mystery was part of the aesthetic attraction of Nature. In contrast, for the two boys who commented on the mysterious Nature of Nature, Nature is mysterious because it is inexplicable.

**Assertion #8**
Religion is a significant factor in many students thinking about Nature. The influence of religious thought shows wide variations of strength. Moreover, a strong religious influence can work either for or against a scientific understanding of Nature.

### Chapter 7: The Environment, Science and Religion

**Assertion #9**
The ninth grade students of this study, on the whole, articulated a high degree of environmental awareness. The relationship of that awareness with science, however, was quite mixed.

**Assertion #10**
Most of the ninth grade students felt obliged to participate in efforts to restore and conserve Nature. This obligation was frequently expressed as a religious or personal, rather than a practical, necessity. The personal the obligation was primarily based on the personal pleasure derived from being out in Nature.

### Chapter 8: High School Science Teachers Talk about Nature

**Assertion #11**
The expected is found. Science teachers show that they know more about science than do their ninth grade students. They speak more about science and they speak in more depth about science, rather than the "name dropping" speech typical amongst the students.

**Assertion #12**
When compared with their ninth grade students, science teachers have a much more focused and less diverse conceptualization of the natural world. As a proportion of their total comments, students speak much more about Nature as the environment, and much more of the aesthetic aspects of Nature and spiritual or religious ideas associated with Nature. In contrast, their teachers are more focused on what one can know about Nature through science.

**Assertion #13**
Most of the teachers spoke about religious or spiritual ideas with respect to Nature, but the teachers were typically more agnostic and more philosophical than their students.

**Assertion #14**
The science teachers like most of their students had a strong aesthetic sense of Nature. They also had a strong sense of Nature as our environment. The biology teachers, however, had a more conservationist view of the environment while the physical science teachers had a more resource view of the environment.

**Assertion #15**
Physical science teachers spoke much more about all that scientists *do* know about Nature and how successful science has been. Biology teachers were also enthusiastic about science. The biology teachers, however, were much less sanguine about science.

# *Chapter 5*
## Science and Conceptualizations of Nature

The research asked to what extent students enjoin scientific knowledge vis-à-vis other domains of knowledge in a discussion about Nature, given that science is unarguably relevant to the topic of Nature; and yet, Nature is a topic that most people do not explicitly associate with science? Moreover, the research asked what are the concepts that appear to have scope and force in the thinking on this topic? The answers from the research are stated in assertions that begin in this chapter. The four assertions discussed in Chapter 5 are assertions specifically related to science.

### Assertion #1

> *The ninth grade students in this study tended to discuss the natural world using concepts from religion, aesthetics, science, and conservationism. A breadth of perspectives, which only sometimes were connected, typically characterized student discussion. A few of the ninth graders demonstrated through their concept of Nature a strong interest in science and voiced specific views about the nature of science. Visible homogeneity among the students (e.g., a classroom of mainly middle class, Anglo students) actually masked substantial variation of thought with regard to Nature, science and science related concepts.*

"Diversity" characterized the students' conversations about Nature. Consider the following excerpt from one of the women students.

> **Patricia**[22]: God created the natural world. It has many characteristics: it's powerful, diverse, changeable, and beautiful... physically and emotionally. The Bible says God created the heavens and Earth, so I think that explains to me what Nature is.... The wonderment of the world increases knowledge through science, but is limited, due to its complexity.... it can also be beautiful in a naturalistic way.... Both views, scientific and religious, try to explain the hard questions, such as the origins of life; in which I believe there is no true answer. Science and religion have distinct roles in our life's teachings. Science teaches

[22] All of the names are gender-sensitive pseudonyms.

> us how to conserve our resources, and how to possibly restore them, while religion teaches us the caring attitudes required to be productive members of the natural world.... People must learn to take the time to enjoy the beauty of Nature, both religiously and scientifically.

Glancing at her narrative one quickly sees religious, aesthetic, and conservationist elements in addition to science elements. Bruce's narrative excerpted below also shows diversity of thought.

> **Bruce:** Nature is complex, but it is orderly and knowable within a 20% or so margin of error. Some things are dependable but not always good like there will always be earthquakes, rain, volcanoes, etc. Most of the natural world can be known through science and the theories that have been developed by science.... Our knowledge is limited by hard questions, such as why does the Earth spin the way it does, what is gravity, and why our planet is solid and not gassy. This mystery and the knowledge we have leads us to a sort of philosophical sort of beauty. Having an open mind allows you to see the beautiful things in Nature, like life in the Sahara Desert. Some people think it is ugly and is a wasteland, but if you think about it, it has an abundant amount of life, which makes it beautiful. It is beautiful in different ways. There is the physical beauty, and there is the emotional or "amazing" beauty.... Some things scientists know about are: weather patterns, El Nino, ozone depletion, and tectonic plate movements.... The natural world is exploited because of us, humans. The earth is in danger because humans are destroying the ozone layer, rain forests and precious land. An incredible thing is that our resources are being exploited and used up, and we need these things to continue our life on this planet. As humans, we have personal and religious obligations to our world to take care of it.

Bruce's ideas about Nature combined ideas about resources, conservation, aesthetics, complexity and order, *and* scientific knowledge. He differed with Patricia in that he had much more to say about science, and much less to say about religion. Also, his view of aesthetics was an intellectual view.

Patricia and Bruce each had a varied but coherent view of Nature in that they tied ideas together. The following excerpt from Holly provides a different example of a diverse conceptualization of Nature.

> **Holly**: The natural world is just there, you know, fish, bugs, dirt, animals, and plants. There are aspects of Nature that have purpose because it was probably created by God, but I am not really religious so I can't explain it. It is very big and complex, like the ocean, which makes it somewhat confusing to know about. There is some order in Nature, but not much.... Cities are not included in the natural world because they are built by unnatural means. Because of these cities, the natural world is exploited; like our resources that we use for medicines, paper and breathing. People need to realize that our resources need to be protected because they are necessities for life. They can be recycled. I do

> not recycle because it is probably not in danger now or during my lifetime, so what's the point?

Patricia and Bruce showed thoughtfulness about Nature and some sophistication in marked contrast to Holly. Holly appeared to have no interest in science, religion, or aesthetics. Her ideas appeared disconnected and she seemed neither interested in, nor concerned about, Nature in general.

Amongst the other students still more variations and similarities appear. Kevin, Alice, and Howard, like Bruce and Patricia, also showed a strong interest in science as they talked about Nature – but in each case emphasis varies as seen in the following two narrative excerpts.

> **Kevin**: I think Nature is very complex. There are unknown parts of Nature and they are confusing to me because there are no real laws controlling them. There is no order.... I think that because Nature is so important to us we need to work to learn more about it. Knowing about Nature makes us feel more at home in it.... There are also knowable parts of Nature. We can learn about Nature through science. There is order to some things and we can base predictions on that.... I want to be a scientist. Science raises many questions about Nature. By trying to answer those questions maybe we can learn to restore some of the changed, damaged parts of Nature. Our environment needs protection for the future. We need to protect the environment by recycling, car-pooling, reducing pollution, conserving trees, etc. The ability to protect requires knowledge.

> **Alice**: I want to be a scientist. Nature is very important to the world of science. Through science we understand many of the patterns in Nature; food webs, weather patterns, how the solar system works, etc. We need to know more about Nature and we keep studying it to find out how things work and to discover ways that different things affect each other. However... it is all people in the world that must act responsibly to help solve the problems we've created in Nature.

Alice and Kevin explicitly say they want to be scientists and emphasized the importance of knowledge. Both talked about environmental issues and using science to promote environmental objectives. In addition Alice talked about patterns in Nature. In contrast, Kevin said, "Nature is very complex.... There is no order" in Nature. Kevin unlike Patricia and Alice does not raise religious issues with regard to Nature, but he did talk about the aesthetics of Nature. Alice spoke about religion but unlike Patricia, Alice indicated that she had accommodated religion to science.

> **Alice**: While I am a Christian, I also believe that science has proved wrong many of the things in the Bible. Yet I do think that there is a purpose for our existence and God is behind it. Science can explain how things work but there are many "why" questions that science doesn't answer.

Howard provided yet another variation on student perspectives supportive of science. In the excerpt below one hears the student who had the most to say about science and with the greatest enthusiasm.

> **Howard**: I think that Nature can be fully known because it is logical.... as time goes on we will understand more and more. Most things about Nature are somewhat orderly or have a pattern to them. Because of this the study of science allows us to explain what is going on in Nature.... The orderliness also lets us predict many things that are going to happen... Nature is very powerful and sometimes it seems chaotic but that is mostly because our knowledge is incomplete and therefore our understanding is limited.... I think that everything can be explained by science.... Nature provides us with many resources. Energy, shelter, food and water all come from Nature. Scientific studies will allow us to use more of Nature to our advantage. Our exploitation has caused pollution.... We are also able to restore it somewhat by conservation efforts. We need to be careful, though, of environmental extremism.

Howard showed a strong inclination toward a scientific, utilitarian view of Nature. He referred only weakly to environmental concerns and he was the one student to suggest that environmental concerns can be taken too far. Howard made no reference to aesthetic or religious aspects of Nature. Rather, he appeared to focus on Nature as logical and orderly and fully amenable to scientific explication. Taken together – Howard, Patricia, Bruce, Alice, and Kevin – are five students who showed an interest in science and whose understanding of Nature involved science yet how very different their perspectives really are.

But, not all students in the study spoke supportively of science. Indeed, other students were very much less sanguine about science. Consider the following excerpt from Art.

> **Art**: Nature is a source of knowledge.... At the present time our know-ledge of the natural world is limited. Many things that we perceive to be complex and confusing because we don't understand them are actually quite simple and orderly. The construction of a spider web, for example, is quite a complicated operation to us but to the spider building the web it is a simple procedure.... It is more important to have a spiritual understanding of Nature than just scientific knowledge. That understanding can't be gained from school. You have to spend time in Nature and learn to feel it. Than you will understand it. The American Indian culture has the kind of understanding for Nature that encourages preservation rather than destruction.... Unfortunately scientists and scientific knowledge are also increasing our tendency to pollute, destroy and clutter up the earth and space. They are trying to destroy it and study it at the same time.

Art is a mystical environmentalist; Howard's opposing bookend. I found Art, like Howard, to be both thoughtful and articulate but his views on Nature and science could not be more different. Howard asserted that "Nature can be fully known because it is logical." Art asserted that one must "learn to feel" Nature. For Howard, Nature seemed only to be a resource for human physical needs. For Art, Nature was also a mental and spiritual resource. Both spoke of Nature as understandable but that is where the similarity ended. Howard was convinced that science will eventually explain all of Nature. In contrast, Art was much more impressed with the knowledge of Nature one achieves by personal experiences with Nature and the type of knowledge traditionally held by Native Americans. Art was as much opposed to science as Howard was supportive.

To summarize this assertion, the students as a group were not found to have a singular conceptualization of Nature. They were not focused on science as was, for example, the teacher Mr. Hess (see Chapter 8, page 75, and Appendix B, page 145). With Howard the one exception, students did not show conceptualizations of Nature dominated by scientific ideas. Along with Howard, four other students showed a strong appreciation for science; but these four, along with the balance of the group, conceptualize Nature as a *composite of a number of different perspectives*: aesthetic, religious, conservationist, and sometimes scientific. In terms of social and economic factors this is a homogeneous group of students. They nonetheless show considerable conceptual variation, for example, a sophisticated Bruce to an unsophisticated Holly, a scientifically inclined Howard to a scientifically alienated Art.

**Assertion #2**

Ninth graders were chosen as an interesting group for this study because they are at an important educational junction point; their elementary schooling lies behind with high school and (for some) college ahead. One sees from the first assertion that the students brought a number of ideas to the conversation about Nature – only one of which was science. This naturally raises an interest in what was said about science. Assertion 2 states:

> *After 9 years of schooling, the level of science integration within everyday thinking remains low among these ninth graders. In their discussions of Nature most volunteer little school knowledge of science. They are aware of school science topics such as the ozone layer, rain forests, or the Big Bang theory. Such topics are voluntarily mentioned, but usually without elaboration even when asked. The ninth graders made few references to general concepts of order or pattern in Nature. Moreover, some students showed signs of a negative perspective on science while others seemed barely to know science exists.*

What should students leaving the K-8 grades know about science? This is not an easy question to answer. The intent of this research is to provoke discussion about expectations for elementary school science by showing how this group of students talked about science and Nature.

Howard and Bruce exemplify the student at the end of elementary education for whom science is very important. Most of the students in this study had much less to say about science in regard to Nature. Holly's most specific comment was that "the natural world is just there, you know, fish, bugs, dirt, animals, and plants". Similarly, Jackie simply offered that,

> Nature... is everything around us like plants and animals... These resources are essential for life and... using them leads to pollution that is destroying our ozone layer.

Even though prompted during the interview, Jackie offered no explanation nor even any examples to go with her statement. She offered nothing further about plants and animals. She named no specific resources nor offered an account of how these resources are essential. Jackie offered no specific examples of pollutants though she did mention ozone destruction.

Betty had somewhat more to say. She seemed to have a more developed sense of what science is and how it works. As seen in this excerpt, science is logical, factual and problem solving.

> **Betty:** Nature is knowable... people know Nature on a scientific or factual basis. Their knowledge is based on facts and can be applied to solving problems as it is logical. There is an order to Nature which we can use to predict some things, weather for example. Ideas about evolution, the ice age, extinction's and global warning can be developed and studied with scientific methods and proofs. Medical cures are another benefit we've gained through factual knowledge.

In contrast to Howard, Betty's statements were relatively simple, and though she listed several science concepts she did not elaborate on any. She did, however, associate an orderliness with Nature that allows a certain amount of prediction, and linked order in Nature and with science. This is of interest since "order in Nature" is an important elementary school science objective (AAAS, 1993). Listening further, one finds that over all Betty has an ambiguous view of Nature and science.

> **Betty:** But people know or understand Nature in two very different ways. Some understand Nature on a religious or spiritual level. They "know" Nature as an emotionally uplifting experience. God and Nature are intermingled in New Age Spirituality. Nature has aspects that can be considered not only to be living and but to also have consciousness.... My understanding of Nature is more scientific and logical than spiritual but there are some aspects of both attitudes in my thinking. Nature is complex and therefore mysterious. It is also powerful. There are many questions that are still unanswered. We don't understand a lot of things in Nature because of its unpredictability. Tornadoes and earthquakes are unpredictable and powerful. Nature is also mysterious because it is living. Things in Nature have a consciousness. Since we are part of Nature and we have feelings, then Nature has feelings. Things in Nature have feelings. Plants, for example, scream when you pick a flower. That is something people don't realize or understand. The consciousness and the beauty of Nature are another type of powerful force.

Though Betty began with orthodox comments about order in Nature and scientific understanding, she ended with startling comments from New Age mysticism, which she acknowledged was an influence in her home. Betty's heterodox views on science, order, and Nature lie in marked contrast to Bruce and Howard. Recall that for Bruce, "Nature is complex, but it is orderly and knowable within a 20% or so margin of error"; and for Howard, "Most things about Nature are somewhat orderly or have a pattern to them.... The orderliness lets us predict many things that are going to happen". These comments are consistent with an orthodox science curriculum. The AAAS noted that, "Science presumes that the things and events in the universe occur in consistent patterns that are comprehensible through careful, systematic study" (1990, p. 2).[23] Howard is explicit about order in Nature. Betty's view of order in Nature is ambiguous at best. Art's view of order is basically a spiritual view. Other students alluded to order in Nature but not as an important attribute of Nature or as a concept fundamental to science.

Four of the students actually seemed much more impressed by *disorder* in Nature. Jackie commented that, "The natural world is incredibly mysterious. There is really no order to what happens. The main mystery is how life came about on this planet". Or, consider the following excerpt:

> **Simon**: I don't really understand a lot of things. I suspect that much of Nature isn't meant to be understood. Because Nature lacks order and is often unpredictable, it is often unexplainable. Animals also do things that we don't understand and can't explain.

Simon was one of three students in this study who never mentioned the word "science". Holly was another. Holly spoke of some order in Nature but also said Nature was "complex" and "confusing". Allen was the third student not to mention science; neither did he speak of any order in Nature. The closest he came to a scien-

---

[23] Natural philosophers and scientists have historically presupposed the existence of order, pattern, and regularity in Nature as a necessity for the development of any meaningful explanation and knowledge of natural phenomenon. In one of our pilot studies, a biology professor volunteered the following:

> "A more scientific impression of Nature, you know, well, I like to look at Nature as being orderly. Although people don't like this word, Nature is orderly but I know there is some discussion about that. My first impressions are always from observations and I can see the order in Nature. Gee whiz, what terrific things! You can't explain everything, but these are what you can explain in Nature that Nature is an orderly process. Volcanic activity is a progression of events that can be predictable. There are patterns. There are patterns from the level of the cell all the way up to the order of a community. You know, so I think it is, it is orderly. Nature is something you can observe and measure so it's very real and you can see the order in it, you can measure it. You can predict it. The fact that you argue about Nature doesn't mean that it's not knowable. You can understand the basic processes of Nature. It is knowable. I don't think that we necessarily know everything, but Nature is knowable. The processes can be worked out. And, yes, Nature is very diverse but it can be orderly and still be very diverse."

Or, as one eminent physicist succinctly wrote, "I do not for a moment believe that the marvelous order science discerns is an order that scientists impose upon the flux of experimental experience" (Polkinghorne, 1996, p. 107). Even Chaos Theory is about recognizing order where there appears to be nothing but chaos, and on the other hand, recognizing the limitations of order in Nature. Hence, this presupposition of natural order is one aspect of a typical scientific worldview (AAAS, 1993).

tific statement was, "Nature is knowable to some extent; like people can recycle and fix the ozone layer by not driving cars and stuff. Then added after some reflection, "But most of [Nature] is not knowable.... For the most part Nature is not orderly and predictable in the sense that nothing stays the same". For Allen, what seems predictable about Nature is that Nature will be better off if people are less intrusive.

It is interesting to note that some adults, with far more life experience and formal education in science, still hold similar views. The following narrative excerpt is from Denise, a college student interviewed for a previous study:

> **Denise**: Manmade things are orderly... but not Nature. Nothing is really solid, nothing is really for sure, you know... an earthquake or something... you're not sure tomorrow's going to come. Things don't always happen the way we think they're going to happen, so it is not orderly the way man-made things are.

Denise is much older than the ninth graders. She has successfully completed science courses through the first year of a college nursing program; yet, she too does not invoke science in her discussion about Nature. As seen in the above excerpt, neither does she think of Nature as having much order. Indeed, she sounds very much like Simon.

As detached from science as Simon, Jackie, Allen, and Holly appeared to be, they were not negative about science. Such was not the case with Art and Paula as seen in the following excerpt.

> **Paula**: Nature is mysterious. I wonder about Nature.... God created the natural world which makes it very mysterious and, for the most part, unexplainable..... Because the earth is God's, humans have no right to mess with it.... With the exception of hippies and white witches, who value spiritual ideals, the emotional values, and the mystery of Nature, man has "doomed" the planet. I don't understand the human world and why people feel the need to study Nature. Studying Nature only causes trouble.... It is a very spiritual world if man's technology would not interfere with it.

Paula's narrative shows a strong sense of the mystery of Nature and spiritual values concerning Nature – and "humans have no right to mess with it." Nature for Paula is as mysterious and unpredictable as it is knowable and predictable for (say) Howard. This asymmetry is reminiscent of Robin Horton's early experiences in Nigeria. As a scientist, he explained to a group of Nigerian university students that he preferred the world of inanimate objects because there one could find "order, regularity, predictability, simplicity" (Horton, 1993, p. 215) things all quite absent in the world of people. He was surprised to discover that his students saw things just the other way around. His perspective on the natural world was, "so totally foreign to their experience that they could not begin to take it in. They just stared" (Horton, 1993, p. 216). As with Horton's Nigerian students, Paula's Nature is not the realm of knowable order. Nature is interesting for Paula, but not in terms of any technical knowledge about Nature. She never explicitly spoke of science but it seems clear that her notion of technology would include science. What we hear is that, "studying Nature only causes trouble." Her views echo Art:

> **Art**: Unfortunately scientists and scientific knowledge are also increasing our tendency to pollute, destroy and clutter up the earth and space. They are trying to destroy it and study it at the same time.

The low level of science talk among these ninth graders is consistent with NAEP assessments of elementary science education achievement (National Research Council, 1996a; Rudner, Song, Treacy, & Pike, 1984) and concern expressed in the science education community (Gardner & Cochran, 1993). The low level of science talk is perhaps excusable for the typical ninth grader especially at the start of the school year, but it is not encouraging when one notes a similar lack of elaboration amongst many older students such as those in Cobern's (1993a) collegiate nursing study. One could take Rutherford's (1987) point of view that elementary science education should focus on a general understanding of what science is about rather than conceptual specifics. In that sense, one would think that while students might not volunteer many scientific details, their years of exposure to science in the elementary grades would give them a sense that Nature is orderly and understandable by scientific methods. That appears not to be the case with this group of ninth graders forcing one to conclude that for most of these students, the science they learned in school had little to do with the natural world they know by personal experience. Moreover, one must agree with Charron (1991, p. 686), "that despite consistent instruction to the contrary, most students assigned science a minor role in their lives." Which is all the more arresting given that several of the ninth graders mentioned an interest in health related careers. However, other researchers have noted a similar disjunction (e.g., Baker & Leary, 1995) in student thinking. In research involving students at similar grade levels, Atwater, Wiggins, and Gardner (1995, p. 672) noted that, "[m]ost of the students planned on a professional career; however, they viewed careers in medicine, engineering, and teaching as not requiring a science background." Indeed, a striking finding in Cobern (1993a, p. 935), "was science's apparent lack of influence on [nursing] students' beliefs about Nature even though these students had been successful in college level science courses."

This disjunction between the students' experience of the world and the world as constructed in the science classroom is symptomatic of what Eger (1992, p. 342) called the "double distancing" between science and Nature that too often takes place in the science classroom. It is all the more unfortunate given the importance the majority of these ninth graders attached to personal experience with Nature, an observation addressed by Assertion 4 (page 50).

**Assertion #3**

In any school science setting, students receive assignment grades, test grades, and course grades. One wonders what evidence of grade differences in science might surface in student conversations? That question is addressed by Assertion 3:

> *Science grade success is not correlated with the concepts these ninth graders typically choose to use in a discussion about Nature. Students with the most science grade success*[24] *have not necessarily grasped fundamental concepts about Nature and science such as the concepts of order and pattern, nor do they necessarily demonstrate a scientifically informed view of Nature. Science grade success does not appear to mean that a student will understand that science is about Nature, nor does having an under-standing of Nature appear to influence school science grades.*

Students like Bruce and Howard perceived order in Nature and that scientific knowledge is to some extent predicated upon order. They have the appearance of Costa's (1995, p. 316) Potential Scientists with "A" grades. As it happens, Bruce and Howard were "B-minus" science students. Betty with her much more ambiguous stance toward order in Nature and scientific knowledge was also a "B-minus" student. Of the three other students (Patricia, Kevin, and Alice) who by their language showed an interest in science, two were "A" students and one was a "B" student. Less surprisingly, Allen and Holly who saw little order in Nature and had *nothing* to say about science were "C" students. In terms of science grade success, however, there was not much difference between Bruce, Howard, Holly, and Allen even though their remarks showed substantial variation.

More surprising were the "A" science students, Patricia, Sally, Liz, Ann, and Kevin. During the interviews each said something positive about science but none choose to speak of Nature as having order or spoke in terms of what (say) the AAAS would consider a scientific worldview.[25] Kevin came close noting that, "there is order to some things and we can base predictions on that. Examples of knowable, predictable things would be states of matter, life cycles, the earth's plates and sometimes the weather". But, he also noted that, "Nature is very complex. There are unknown parts of Nature and they are confusing to me because there are no real laws controlling them. There is no order". Indeed, Kevin's interest in science had little to do with the traditional notion of fathoming the depths of Nature through science, and everything to do with his environmentalist perspective, as seen in this excerpt:

> **Kevin**: The resources in our environment are a necessity to us for our survival. But our growing need has led to exploitation due to people's lack of caring.... I want to be a scientist. Science raises many questions about Nature. By trying to answer those questions maybe we can learn to restore some of the changed,

[24] The notion of "grade success" was suggested by the teacher/researchers in this study, recognizing that grades do not always reflect knowledge and understanding of the course objectives. We simply checked student records to see if their past and present science grades had typically been "As and Bs," "Cs," or less.

[25] The notion that there is a single scientific worldview or single nature of science or single philosophy of science employed by scientists is at best a problematic notion. Notions of this sort, however, are readily found in most science curriculum and framework documents, which is why I have used the notion of a scientific worldview in this study. The results of this study, on the other hand, suggest that setting some notion of a "scientific worldview" or "scientific outlook" as a key curriculum goal is itself problematic. The concept of "order" in Nature is similarly problematic but I think ultimately of much more importance.

> damaged parts of Nature.... We need to protect the environment... The ability to protect requires knowledge.

Kevin was not the only student holding this type of perspective.

> **Sally**: The natural world is somewhat knowable through science and religion. It is too big to be entirely explained. For example, how can you be sure that an animal is truly extinct if you can't explore all areas of the world? Science and scientists help us to know some of the natural world because things can be predicted, like animal behavior. The predictableness allows us to answer how things work, but we will never really know why things work. Why is Nature here? What is the purpose? How did life form? Some things are unpredictable like hurricanes, tornadoes, and volcanoes, which make Nature dangerous at times. Science can teach us how to be better conservationists through research and technology so we can avoid pollution which ruins Nature.

Similarly, Ann noted that some events in Nature are predictable but that Nature can also be confusing.

> **Ann**: Nature is knowable but the questions I ask about Nature make me think that Nature is sometimes very confusing. It is also changeable. There are some things like the weather that we can predict but other things are not predictable. Some things, like earthquakes, can be dangerous because of their unpredictability. We can learn to understand many things about Nature through personal experience, school and science.

Liz is yet another of the "A" students who does not speak of scientific knowledge grounded in the orderliness of Nature. She is quite optimistic about what can be expected from science but that expectation is bounded by her aesthetic and religious views.

> **Liz:** The natural world is all the animals and the things around us. It also includes the environment and how they interact with it. The natural world also consists of ideas, why animals do certain things, their purpose, and what they think. It is the work of God. [Nature's] purpose is to help us live and enjoy the things (pause) the aesthetics (pause) it provides us. Everything happens for a purpose. The natural world is knowable by means of education through science and by learning through personal experiences. Eventually we will probably be able to know most things about the natural world. However, some things will be kept a mystery because not all things are meant to be known. Science tends to teach the how and what questions about the natural world and religion hints at the why questions, somewhat. Before it can be knowable to someone, that person must care about the natural world. Lack of care not only hinders your personal thought, it sometimes leads to exploitation of natural resources and natural environments like the rainforests.

Of course, as Ann commented, Nature *can* be confusing – even for the best of scientists. What is missing from the remarks of these "A" students, is any conviction that there is an underlying order in Nature that makes science possible. Einstein's thought that the only thing incomprehensible about Nature is that it is comprehensible is not evident here. Not that this keeps these students from appreciating science or getting good science grades, which it clearly does not. Rather, it is an indication that at this point in their lives their conception of scientific knowledge and the natural world does not rest on the Western tradition of Nature as an orderly system. This group of students was much more likely to be interested in science for environmental reasons – a point that will be addressed in Chapter 8.

What would one want to hear from science students, especially science students who have done well at school science? Documents such as *Project 2061* and the *National Standards for Science Education* suggest something similar to the comment made by a college student sitting for the same type of interview. Clara commented, "I'm a science major and... it's everything I've been taught that everything has an order to it... a system to it" (Cobern, 1993a, p. 942). Clara says she learned about order in Nature through her science classes, there are questions that beg asking. Does science instruction really help students discover order in Nature as Clara claims and as curriculum policy suggests should happen? Is it possible that the opposite is the case, that is, that a pre existing sense of orderliness precedes a certain sense about science? Or is it possible that some (many?) students simply fail to see much connection between the *scientific* order of the science curriculum and what they see in the real world?

In summary, one might think that students with grade success in science would have grasped an important assumption in science, that is, that science is predicated upon an assumed orderliness in Nature. That does not appear to be the case with this group of ninth graders. Moreover, even among the best students one gets little sense of a scientific worldview as described by the AAAS (1993); and it is of considerable interest that grade-successful students after nine years of school have so little to say about either a major assumption of science (i.e., order in Nature) or the processes of science. One is reminded of Costa's (1993) description of school science as *a rite of passage* by which society helps some students make the transition from "school student to scientist or member of the scientific community" (Costa, 1993, p. 654). One can imagine how this description fits Howard and perhaps Bruce. One can see them continuing in school science and beyond. As for the other students, if school science is a rite of passage to the community of science and scientists they seem only dimly aware of that. They could just as easily be playing by "Fatima's Rules" (Larson, 1995, p. 8). In an interpretive study of high school chemistry students, Larson (1995) met a young women, Fatima, who got good and honest grades in her chemistry course but not by a method that would have pleased her teacher had he known. These rules include, "Don't read the book.... Ask the teacher for help as soon as you're stuck" (Larson, 1995, p. 8). The rules work but as another student admitted,

> I get an 'A' because I do my work, but as for like, if you could grade me on how much I *know*, I'd get an 'F'... I don't understand [chemistry]. (Larson, 1995, p. 12)

Nor, one might add, is it likely that the student's school science experiences have had much impact on the student's everyday thinking about the world in which she lives.

**Assertion #4**

The last assertion to be considered in this chapter is an assertion about personal experiences in Nature. It is perhaps not surprising that when given the opportunity to speak freely, many students freely spoke about the things they liked to do. Assertion 4 reads:

> *Most of the ninth graders, regardless of school science grade success, attached considerable importance to their personal experiences with Nature.*

Many of the student narratives show emotionally buoyant responses to Nature that cannot easily be associated with anything related to school science or to school. These responses were more personal. The following comments are typical.

> **Simon**: I really enjoy being out in Nature. It gives me good feelings. I like walking around, climbing mountains, watching a deer drink out of a river and things like that. I think about Nature and you could say I'm in touch with Nature.
>
> **Howard**: I find Nature to be peaceful when I'm hiking up a mountain or something like that. But I also find it peaceful when I'm just walking around at night sometimes.
>
> **Samantha**: The pleasure I get from being in Nature is very important to me. I spend a lot of time in Nature. I'd be pretty bored if I didn't have it.
>
> **Ann**: I like to go where you can't see any influence by man. When I'm out in Nature I feel calm and peaceful.... Sometimes, when I'm out in Nature and I have time to think...
>
> **Kevin:** I live in a "natural" area. Being in Nature is important to me. I can see and feel it so I know it exists. I enjoy the beauties of Nature, the animals, mountains etc. It supports my sense of self.
>
> **Sally:** Sometimes it is nice to go somewhere humans have never been. For example, I like going to a totally natural place like a grassy tree area where I can read and just ponder things. It would be really neat to go in a capsule under the ocean and just be absorbed by the beauty. I guess my mom instilled this value of beauty in me.
>
> **Liz**: I live out in the desert where I can enjoy looking and thinking about the animals that live there. I get peace from that.

Not all students mentioned such personal experiences, but of the group of students, only Holly was indifferent to Nature. Of the students who through their remarks showed themselves to be more science-oriented, only one described his aesthetic sense of Nature in intellectual terms that could be associated with science. Though aesthetic[26] remarks and personal experiences were sometimes associated with religious and environmental views, they were *not* typically associated with either school or science.

In the first four assertions of this study, gender issues do not arise. In none of these assertions were there indications that the boys and girls of the study were different. They were not noticeably different in the areas described by the first four assertions. The next chapter examines gender associations where they seemed most likely to occur.

Before leaving this chapter, there are a few more things that need to be said about these ninth grade students and their knowledge of science. As noted in this chapter and in Appendix A (page 127), some of these students have done very well at school – including science – and some have done quite poorly. Their science teachers attest to the fact, however, that all the students, including the weakest student in the group, know some science. Moreover, most of the students also know that scientific knowledge is knowledge about Nature. After the series of individual interviews, I met with the students in two groups of eight. We talked about school and about the interviews; and I took the opportunity to comment that I wondered why so little had been said about science during the interviews. Howard quickly protested that he indeed had said much about science, which I acknowledged as true. Several of the other students had a different comment. They said, "but you didn't ask us about science." To this I responded, "but I didn't ask you about any subject yet you volunteered ideas about beauty, religion, the environment, but not much about science." They agreed that my statement was accurate but could offer no explanation for why science did not come readily to mind during the interviews. My explanation is that the ideas most important to the students were the ideas that came to mind. For Howard, the one the teachers call a "science information gatherer," science immediately came to mind showing its importance for him. It seems clear that for other students, science was much lower on the "importance scale" vis-à-vis other ideas and for some it did not even register. So, what was it about their school science experience through eight grades that science should have such a minimal impact on their *everyday* thinking?

[26] Assertion #6, discussed in the next chapter (page 56), is a closely related assertion specifically addressing aesthetics and Nature.

# *Chapter 6*
## Gender and Conceptualizations of Nature

Given the attention that gender issues receive in science education (Baker, 1998; Kahle & Meece, 1994) and feminist scholarship on the social construction of Nature (Merchant, 1989), upon examining the student narratives one immediately looks for possible gender differences. Do the boys talk about science more than do the girls? Do the girls talk about aesthetics and religion more than do the boys? Are the girls more environmentally sensitive than are the boys? These questions and others are implicitly suggested by the literature. They are addressed in the assertions that follow.

Figures 8a, b and c roughly summarize the diversity of thought that was first discussed under Assertion 1, and to show the diversity that was found broken out by gender. A student is included in a category if he or she commented in some substantial way relevant to the category.[27] The magnitudes of difference shown in Figures 8b and 8c are small with one exception. It is particularly clear in Figure 8c that religion was a more frequent topic of conversation with the girls than with the boys. This observation is dealt with in Assertion 8. However, it is important not to over interpret the differences seen in Figure 8c given the small size of this study. Therefore, I have not taken the other differences as particularly meaningful.[28]

### Assertion 5

The first four assertions were treated without reference to gender because in the development of those assertions there never appeared to be any significant gender factor. That pattern continued as the research specifically examined questions about science and the environment.

> *The ninth grade boys and girls of this study appeared to have few differences in the way they conceptualized Nature. Both girls and boys regarded science as an important way of knowing about Nature – actually more*

[27] In other words, to be included in a category a student had to make more than a passing or off-handed remark.

[28] Of course, such differences could be examined statistically but our group did not feel that a quantitative analysis at this point would serve our interests; nor was the study set up to enable a quantitative analysis.

| | A<br>Maj | B<br>Sci+ | C<br>Sci- | D<br>Mys | E<br>Aes | F<br>Rel | G<br>Env | H<br>Per |
|---|---|---|---|---|---|---|---|---|
| **10 girls** | | | | | | | | |
| Alice | X | X | | | X | X | X | |
| Ann | X | X | | | X | X | X | X |
| Betty | | X | | | | X | X | |
| Jackie | | X | | X | | X | X | X[29] |
| Liz | | X | | X | X | X | X | X |
| Holly | | | | X | | | | |
| Patricia | X | X | | | X | X | X | |
| Paula | | | X | X | X | X | X | X |
| Sally | X | X | | X | X | X | X | |
| Samantha | | X | | | X | X | X | X |
| **Girls Freq.** | **4** | **8** | **1** | **5** | **7** | **9** | **9** | **5** |
| **6 boys** | | | | | | | | |
| Allen | | | | | | | X | |
| Art | | | X | X | X | X | X | X |
| Bruce | X | X | | | X | | X | |
| Howard | X | X | | | X | | | X |
| Kevin | X | X | | | X | | X | X |
| Simon | | | | X | X | | X | X |
| **Boys Freq.** | **3** | **3** | **1** | **2** | **5** | **1** | **5** | **4** |
| **Total:** | **7** | **11** | **2** | **7** | **12** | **10** | **14** | **8** |

| | Maj | Sci+ | Sci- | Mys | Aes | Rel | Env | Per |
|---|---|---|---|---|---|---|---|---|
| **Girls %** | **0.4*** | **0.8** | **0.1** | **0.5** | **0.7** | **0.9** | **0.9** | **0.5** |
| **Boys %** | **0.5** | **0.5** | **0.2** | **0.3** | **0.8** | **0.2** | **0.8** | **0.7** |
| **Δ** | **0.1** | **0.3** | **0.1** | **0.2** | **0.1** | **0.7** | **0.1** | **0.2** |

A. Student expressed interest in a science or science-related college major or career.
B. Student commented *positively* about *science.*
C. Student commented *negatively* about *science.*
D. Student commented on how Nature is *unknowable* or that it is *mysterious.*
E. Student commented on the importance of *aesthetic* aspects of Nature.
F. Student commented on religion and *religious* aspects of Nature.
G. Student showed personal commitment to an *environmentalist* view of Nature.
H. Student commented on personal experiences with Nature.

* The numbers in these rows are simple averages.

Figure 8a. The Representation of Ideas across Students by Gender

[29] During her primary interview, Jackie did not speak of personal activities in Nature or personal feelings about Nature. However, in the interview in which Jackie was asked to talk about herself, her family and friends, and activities, she expressed a clear interest in camping and other such activities.

Figure 8b. Percent of Boys and Girls in Each Category

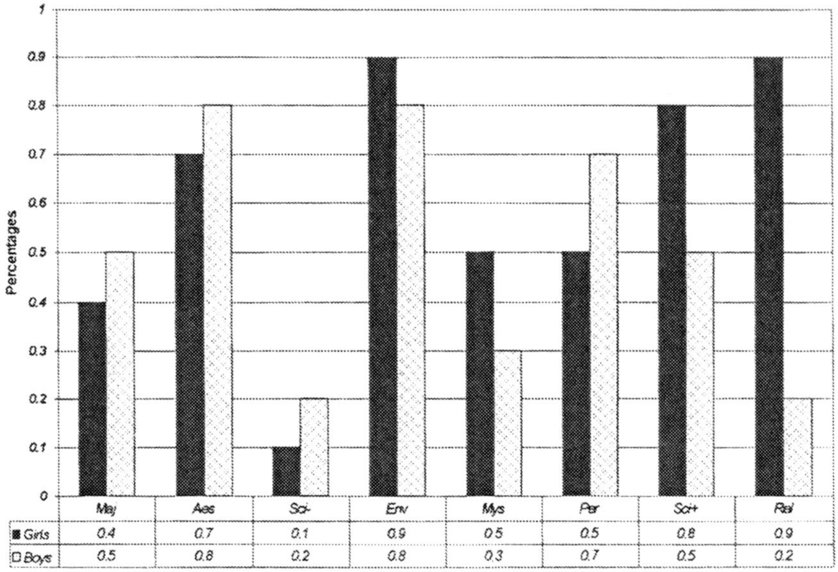


Figure 8c. Percentage Differences between Boys and Girls

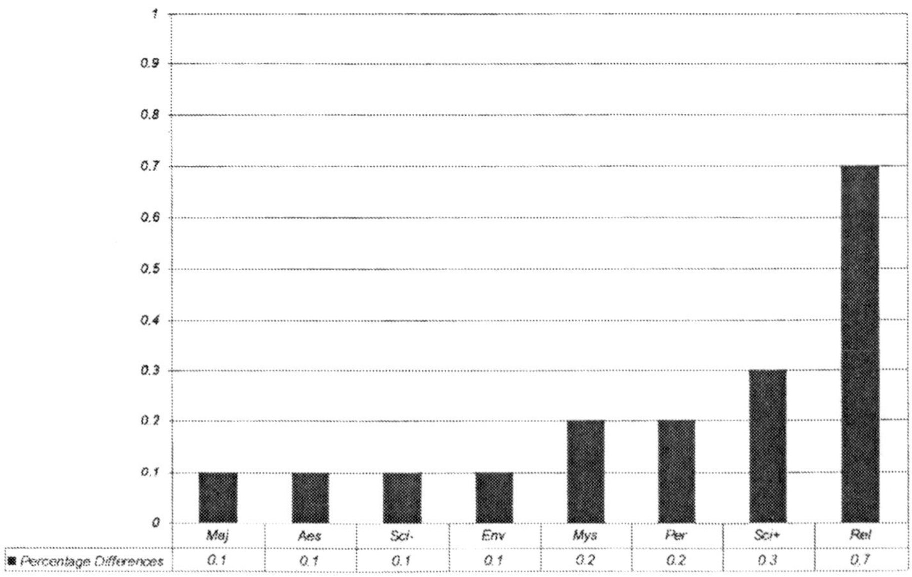

*girls than boys. The ninth grade girls were no more likely to hold caring, environmental views of Nature than were the boys although the literature suggests that these are more typically feminine perspectives.*

For this group of ninth graders, boys are no more interested in science and science careers than are the girls. For example,

> **Alice**: I want to be a scientist. Nature is very important to the world of science. Through science we understand many of the patterns in Nature; food webs, weather patterns, how the solar system works, etc. We need to know more about Nature and we keep studying it to find out how things work and to discover ways that different things affect each other.
>
> **Kevin**: We can learn about Nature through science.... I want to be a scientist. Science raises many questions about Nature. By trying to answer those questions maybe we can learn to restore some of the changed, damaged parts of Nature.
>
> **Bruce**: Most of the natural world can be known through science and the theories that have been developed by science. Science enables us to predict, to some extend everything such as weather, volcanoes, earthquakes and earth movement.
>
> **Sally**: Science and scientists help us to know some of the natural world because things can be predicted, like animal behavior. The predictableness allows us to answer how things work...

Four of ten girls and three of six boys expressed some interest in science or science-related careers. All seven of these students also commented positively about science.

Of the sixteen students, only two spoke negatively about science (explicitly or implicitly), one girl and one boy.

> **Art**: Unfortunately, scientists and scientific knowledge are also increasing our tendency to pollute, destroy and clutter up the earth and space. They are trying to destroy it and study it at the same time.
>
> **Paula**: I don't understand the human world and why people feel the need to study Nature. Studying Nature only causes trouble. It creates more technology and curiosity, which leads to the exploitation of the land.

On the other hand, if there is a true difference in positive attitudes about science among this group of students, it slightly favors the girls not the boys (see Figure 8b).

Similarly, 14 of 16 students showed an interest in the environment. The issue of environmentalism is addressed in the next chapter. Suffice it to say here, no gender differences were found with regard to environmentalism. Both the boys and girls of this group were quite interested in, and sensitive to, environmental concerns.

**Assertion 6**

Assertion 4 noted the important role of personal experience with Nature, a role that cut across gender lines. The issue of gender with regard to recognition of aesthetic dimensions of Nature, however, showed an interesting complexity that is examined in Assertions 6 and 7. Assertion 6 reads:

> *Most of the ninth graders, whether boys or girls, gave some form of aesthetic response to Nature, which took one of three (sometimes overlapping) forms: a conceptual perspective form, a religious (or spiritual) emotional form, and a non-religious emotional form.*

Twelve of sixteen students spoke aesthetically about Nature in one way or another. Some of the students gave conceptually oriented aesthetic views. They spoke of a sense of beauty in Nature grounded in conceptual knowledge. Assertions 1, 2, and 3 address the conceptual knowledge the students used in their discussions of Nature. In the excerpts below from two of the most science-inclined students, one sees how some students specifically tie that knowledge to their aesthetic ideas about Nature. For example,

> **Bruce**: This mystery and the knowledge we have leads us to a sort of philosophical sort of beauty. Having an open mind allows you to see the beautiful things in Nature, like life in the Sahara Desert. Some people think it is ugly and a wasteland, but if you think about it, it has an abundant amount of life, which makes it beautiful. It is beautiful in different ways.
>
> **Howard**: I do enjoy being out in Nature and thinking about different aspects of it. There is a beauty in all the ideas that there are in Nature to learn about.

Other students expressed their aesthetic and emotional views of Nature grounded in religious belief. Assertion 8 notes that more girls than boys used religious language in their discussion of Nature. That difference holds here as well where one sees more girls than boys drawing upon religious themes in their comments about the aesthetics of Nature. For example,

> **Patricia**: God created the natural world. It has many characteristics. It's powerful, diverse, changeable, and physically and emotionally beautiful.
>
> **Liz**: [Nature] is the work of God. Its purpose is to help us live and enjoy the beautiful things it provides us.... The natural influences the way we think. Just thinking about the natural world gives us peace and energy by knowing that the animals are okay and to see that we are not the only organisms living here. I live out in the desert where I can enjoy looking and thinking about the animals that live there. I get peace from that.
>
> **Sally**: Knowing that the natural world was given to us by God gives me a wonderful and uplifting feeling.... [Nature] is a beautiful place, not only physically but emotionally.

> **Art**: There is a spiritual aspect to Nature for many people. I find it quite beautiful, especially when I am looking out at a part of the natural world that is untouched by man. I appreciate the diversity of plants and animals in Nature. Animals are very important to me. I can feel things through animals. I enjoy watching them and learning about them. I understand why Nature is sacred to some groups of people. It is sacred to me.

Ann's narrative is of particular interest because she discussed her aesthetic and emotional sense of Nature in both religious and conceptual terms with the two being mutually supportive of her aesthetic views. In the following excerpt, conceptual ideas are indicated by her statement that "I'd like to find answers", which other parts of her narrative show to be a reference to conceptual understanding of Nature.

> **Ann**: To me, Nature is beautiful and pure because it is God's creation.... Nature provides both aesthetic and emotional pleasure and I need it for self-renewal. I like to go where you can't see any influence by man. When I'm out in Nature, I feel calm and peaceful. It is a spiritual feeling and it helps me understand myself. I also get a spiritual feeling from Nature. Sometimes, when I'm out in Nature and I have time to think, I start to wonder about things. This leads me to ask questions that I'd like to find answers to. The pleasure I get from Nature is enhanced by the mysteries I see in it.

In contrast to the above students, however, most of the aesthetic and emotional views that students expressed were linked to neither conceptual nor religious thinking. The views were expressed more simply.

> **Paula**: Nature is mysterious. I wonder about Nature. I would enjoy living in the mountains where the ground has been untouched by humans so I could appreciate the beauty and purity of the natural world.

> **Simon**: I really enjoy being out in Nature. It gives me good feelings. I like walking around, climbing mountains, watching a deer drink out of a river and things like that. I think about Nature and you could say I'm in touch with Nature. Though I understand only a little about it, I like the mystery of not understanding everything. It adds to the beauty. Nature can be peaceful, with calm breezes, lots of nice trees and no trash.

> **Kevin**: I live in a natural area. Being in Nature is important to me. I can see and feel it so I know it exists. I enjoy the beauties of Nature, the animals, mountains etc. It supports my sense of self.

The students who spoke most strongly about the beauty or emotional appeal of Nature often spoke of personal experiences of a recreational type with Nature (see Assertion 4, page 50). Simon, for example, spoke of hiking and climbing. Ann referred to time spent out in Nature. Moreover, several of these students live in rural areas. Liz tells us that she lives in the desert. Kevin says he lives in a natural area. It seems likely – and of little surprise – that home location and recreational activities would be formative elements in a person's thinking about Nature

Clearly, these ninth graders are an aesthetically oriented group of students and there is reason to think that these characteristics will stay with the students. After all, quite similar views were noted among college students in a previous study (Cobern, 1993a, p. 946):

> **Georgia**: I don't feel that Nature is always beautiful or pleasant... it's not always delightful... but it's a place that you can always feel spiritual... close to God and close to your own feelings. It's a special place that God made. It's relaxing and helps you put things in perspective.
>
> **Jackie**: There are a lot of unknown things about Nature. I wonder about all of them. I wonder about the little insects in the forest and things like that so it's a mystery to me how it's things all fit together and works together... I get this feeling it's like a special place... it's kind of holy... Nature just gives me a very special feeling, I mean, when I'm like out in the woods... the beauty and the mysteriousness of it all and it's sort of religious to me. I mean, it's like a special place. I'm not a set person in one religion but Nature does make me feel that there is a God when I see how things are created and what is happening.

As noted in that study, the aesthetic gap between students and the traditional science curriculum with respect to how one can understand the natural world should not go unexamined on the assumption that such differences are insignificant with respect to science education. One should at least seek to determine what significance this assumption may be obscuring. Furthermore there is perhaps more value for science education in the various programs that provide urban students with Nature experiences then is currently recognized (e.g., see Grumbling et. al., 1991).

**Assertion 7**

Although no apparent differences were noted between the boys and girls of this study with respect to aesthetics and Nature, within the complexity of their views was an interesting related difference.

> *For some of the ninth grade girls, the mysteriousness of Nature was an aesthetic element in their conceptualizations of Nature. This was true for only one of the ninth grade boys. For some of the ninth grade girls, the mysteriousness of Nature was also part of a sense that Nature can be inexplicable. In contrast, the boys tended to view knowledge about Nature in terms that are much more positivistic.*

Mystery and mysteriousness can refer to the inexplicable, but it can also refer to an aesthetic element of something that is not fully known or explicated. From this element of aesthetics, one need not infer that things are unknowable. Knowledge may well come later, but the mystery of the moment is aesthetic. Several of the ninth grade girls in this study expressed this idea as part of their conceptualizations of Nature. In the mystery of Nature, they see beauty.

> **Ann**: When I'm out in Nature, I feel calm and peaceful. It is a spiritual feeling and it helps me understand myself. I also get a spiritual feeling from Nature. Sometimes, when I'm out in Nature and I have time to think, I start to wonder about things. This leads me to ask questions that I'd like to find answers to. The pleasure I get from Nature is enhanced by the mysteries I see in it.

> **Samantha**: Even though Nature is complex and hard to understand I think Nature is inspirational. It has a powerful effect on me. Words like beautiful, powerful, pure and peaceful come to my mind when I think about Nature.

> **Patricia:** [Nature is] powerful, diverse, changeable, and beautiful (pause) physically and emotionally. Nature... is anything made by God, all the plants and animals on earth and the entire solar system. The Natural World is very mysterious to me, I wonder about many things in Nature. Something I wonder about is, what is way out in the universe, perhaps another earth? ... You don't have to know about things in Nature to recognize that they are beautiful and sometimes pure.

Of the boys, only Bruce spoke of mysteriousness in Nature that he interpreted as an aesthetic element of Nature.

> **Bruce**: Our knowledge is limited by hard questions, such as why does the earth spin the way it does, what is gravity, and why is our planet solid and not gassy. This mystery and the knowledge we have, leads us to a sort of philosophical sort of beauty.

Earlier I described a sense of beauty that was grounded conceptually. I gave Bruce as an example of one who saw Nature this way. Here again we see something similar. Bruce sees a mysteriousness in Nature that contributes to his sense that Nature is aesthetic, but he called this a "philosophical sort of beauty" – something that none of the girls did.

For some of the other ninth grade girls, mysteriousness and questions about Nature go beyond aesthetics. For these girls there was the sense that Nature can be very difficult to understand and perhaps, in some very deep way, even inexplicable. Ann is a good example of a student who has a high regard for what can be known about Nature, but then says:

> **Ann**: Nature is knowable but the questions I ask about Nature make me think that Nature is sometimes very confusing. It is also changeable. There are some things like the weather that we can predict but other things are not predictable. Some things, like earthquakes, can be dangerous because of their unpredictability.

Jackie combines the same sense of ambiguity,

> **Jackie**: There are so many things to think about when dealing with Nature. It is very mysterious; how did the earth originate, how does a baby form in the mother's womb, which makes it very confusing and unpredictable. The lack of

> predictableness, can be dangerous, such as hurricanes and earthquakes. I don't mean to say that it is totally mysterious. There are some things that can be knowable, like medical science.

Similarly, Patricia seems to say that we both can and cannot know things about Nature.

> **Patricia**: The Natural World is very mysterious to me, I wonder about many things in Nature. Something I wonder about is, what is way out in the universe, perhaps another earth? Even though Nature is mysterious, everything is knowable but maybe not in the near future.... science and religion, try to explain the hard questions such as the origin of life, in which I believe there is no true answer.

For other girls, there is a deep sense of the inexplicable in Nature.

> **Paula**: God created the natural, which makes it very mysterious and, for the most part, is unexplainable. God intended it to be here for a purpose, which is only known by Him. Because it is God's, humans have no right to mess with it. Even with the best technology and scientists we will probably not every fully understand Nature.

> **Sally**: The natural world is somewhat knowable through science and religion. It is too big to be entirely explained, for example, how can you be sure that an animal is truly extinct if you can't explore all areas of the world. Science and scientists help us to know some of the natural world because things can be predicted, like animal behavior. The predictableness allows us to answer how things work, but we will never really know why things work: Why is Nature here? What is the purpose? Or, How did life form? Some things are unpredictable like hurricanes, tornadoes, and volcanoes, which make Nature dangerous at times.

> **Liz**: Eventually we will probably be able to know most things about the natural world. However, some things will be kept a mystery because not all things are meant to be known. Science tends to teach the how and what questions about the natural world and religion hints at the why questions somewhat.

This ambiguous sense about Nature, this sense of mystery was largely absent from the boys conversations about Nature. There was the exception; Simon, who confessed,

> **Simon**: Although I've thought a little bit about the natural world, I don't really understand a lot of things. I suspect that much of Nature isn't meant to be understood. Because Nature lacks order and is often unpredictable, it is often unexplainable.

And, there was Art with his iconoclastic views,

> **Art**: At the present time our knowledge of the natural world is limited. Many things that we perceive to be complex and confusing because we don't understand them are actually quite simple and orderly. The construction of a spider web, for example, is quite a complicated operation to us but to the spider, building the web it is a simple procedure.

For Art, gaining knowledge of Nature – to thus limit the unknown – usually means gaining "the spider's" perspective. The other boys, in contrast to the girls and to Simon and Art, were much more positivistic about Nature when they addressed this point.

> **Howard**: I think that Nature can be fully known because it is logical. We don't know or understand all of it yet but as time goes on we will under-stand more and more. Most things about Nature are somewhat orderly or have a pattern to them. Because of this, the study of science allows us to explain what is going on in Nature.

> **Bruce**: Nature is complex, but it is orderly and knowable within a 20% or so margin of error. Most of the natural world can be known through science and the theories that have been developed by science.... Questions about how the natural world works is explainable through science. Chemists and astrophysicists can come up with theories based on the order and the predictableness of the phenomena in the world.

Kevin's views are a little different. At first glance, he seems to take the same ambiguous stance that Patricia took,

> **Kevin**: I think Nature is very complex. There are unknown parts of Nature and they are confusing to me... There are also knowable parts of Nature.

Later he added,

> **Kevin**: We can learn about Nature through science.... I want to be a scientist. Science raises many questions about Nature. By trying to answer those questions...

The questions he refers to are environmental questions, and the important point is that he indicates no doubts but that science can indeed answer the questions.

In summary, the ninth grade girls of this study were much more inclined to see mystery in Nature than were the boys. This sense of mystery, however, was not problematic but typically an aspect of Nature's aesthetic appeal. As we shall see in Chapter 8, in this regard the girls were closer to their biology teachers (both males) than they were to the physical science teachers (one male and one female). The boys of this study, on the other hand, are more similar to the physical science teachers.

**Assertion 8**

The discussion thus far has seen the topic of religion arise numerous times. There is no doubt that religion was an important factor in many of these students' lives as stated in the following assertion.

> *Religion is a significant factor in many of the ninth graders' thinking about Nature, more so for the girls than for the boys. The influence of religious thought varies and can work either for or against a scientific understanding of Nature.*

All but two of the students in the study made at least passing comments about God, religion, or spirituality with regard to Nature. Nine of ten girls spoke seriously of the topic, but only two of six boys did so. For students like Jackie, Liz, Samantha and Ann, religion constitutes an important reality in life and this shows in conversation about Nature.

> **Jackie**:[30] The Natural World is everything that God created and therefore has a purpose. The Indian culture believes that this land is sacred and should be taken care of.

> **Liz**: It is the work of God. Its purpose is to help us live and enjoy the things – the beauty it provides us. Everything happens for a purpose.

> **Samantha**: My feelings about Nature include religious feelings too. Sometimes when I think about Nature, I also think about God. These are my first thoughts when you say the word 'Nature'.... Nature comes from God...

> **Ann**: Nature is... God's creation.... I also get a spiritual feeling from Nature. Sometimes, when I'm out in Nature and I have time to think, I start to wonder about things. This leads me to ask questions that I'd like to find answers to.

For Samantha and Ann religion appears also to support their interests in science and their scientific views of Nature; however, the religious beliefs and knowledge about Nature are articulated differently by other students. In the excerpt below, one sees that Alice acknowledges a personal religious commitment, on the one hand. Then on the other, she notes a contradiction between her religion, as she understands it, and her scientific understanding of Nature.

> **Alice**: I want to be a scientist. Nature is very important to the world of science. Through science we understand many of the patterns in Nature; food webs, weather patterns, how the solar system works, etc. We need to know more about Nature and we keep studying it to find out how things work and to discover ways that different things affect each other.... I'm not sure about the connection between God and Nature. While I am a Christian, I also believe that science has proved wrong many of the things in the Bible. Yet, I do think that

---

[30] Note that Jackie is Native American.

> there is a purpose for our existence and God is behind it. Science can explain how things work but there are many why questions that science doesn't answer.

For Alice, religion and science serve different purposes. They address the questions why and how, respectively; but for Alice science and religion can also be in conflict. In those conflicts, she appears to side with her understanding science.

Two other girls, rather than finding conflict, assigned science and religion to different tasks with respect to Nature. Betty does this from the perspective of New Age religious belief.

> **Betty**: But people know or understand Nature in two very different ways. Some understand Nature on a religious or spiritual level. They know Nature as an emotionally uplifting experience. God and Nature are inter-mingled in New Age spirituality. Nature has aspects that can be considered not only to be living and but to also have consciousness. Other people know Nature on a scientific or factual basis.... My understanding of Nature is more scientific and logical than spiritual but there are some aspects of both attitudes in my thinking.

Patricia is another student who assigns different roles to science and religion, but Patricia does so from a more traditional religious background.

> **Patricia**: God created the Natural World. In terms of religion, the Natural World is knowable because we have faith in the purpose of it, even though we don't necessarily know it. There is some conflict between the Bible's teachings and views of scientists and environmentalists. Both worlds, science and religion, try to explain the hard questions such as the origin of life, in which I believe there is no true answer. Science and religion have distinct roles in our life teachings. Science teaches us how to conserve our resources and how to possibly restore them. Religion teaches us the caring attitudes required to be productive members of the natural world.

For other students, conflicts between scientific and religious views can pose much more confusing and difficult dilemmas. Paula is an interesting student in this respect. To begin with, she neatly divides her world into Nature with people and Nature without people; and she clearly viewed Nature without the corrupting influence of people as ideal. Paula has a primitivist, anti-technical outlook on life and speaks approvingly of pre modern life. Her views include a strong spiritual element.

> **Paula**: The natural world was in existence before the dawning of man. God created the natural, which makes it very mysterious and, for the most part, is unexplainable. God intended it to be here for a purpose, which is only known by him. Because it is God's, humans have no right to mess with it. Even with the best technology and scientists we will probably not every fully understand Nature. When man entered this planet, he destroyed its purity, beauty, and power. With the exception of hippies, who value the spiritual ideals, the emotional values, and the mystery of Nature, man has doomed the planet. I don't understand the human world and why people feel the need to study Nature.

> Studying Nature only causes trouble. It creates more technology and curiosity, which leads to the exploitation of the land.

In contrast, only one of six boys made anything more than a passing remark about religion and Nature. Howard, who is known by his teachers to be very involved with church and said as much in his interview, offered very little.

> **Howard**: I am a religious person but I also try to take things as they appear to me so I don't believe that everything in Nature has a spiritual side to it.

Similarly, Simon would only say,

> **Simon**: I also have some religious feelings about Nature. Not necessarily those of any one particular religious group. I do think that some God created the earth. This confuses me also. I'm not entirely sure of my beliefs but I do think that a god created the earth.

The one person who did have more to say was Art, who is much taken by a Native American perspective, though he gave no indication of any personal religious practices.

> **Art**: You have to spend time in Nature and learn to feel it. Then you will understand it. There is a spiritual aspect to Nature to many people.... It is sacred to me. The American Indian culture has the kind of understanding for Nature that encourages preservation rather than destruction.... They do not think of themselves as superior beings... Nature is something felt!

Assertion 1 spoke of the diversity of views held by this group of ninth graders. One can now see that this diversity includes religious thought – more so for the girls than for the boys. The students are not unlike adults who also have been shown to have various interpretations of religion and Nature (Loving & Foster, 1998; Roth & Alexander, 1997). Amongst the ninth graders of this study, however, there was no one such as Carla in Cobern (1993a, p. 940) who expressed atheist beliefs.

> **Carla**: I think of Nature as everything that pertains to the planet. It arises from the planet. That involves the air we breathe, the oceans, the earth itself, the land, the living organisms that inhabit it... animals... flora... fauna... minerals... weather phenomena... I think it's everything. A religious person would say Nature is there for a purpose. God put it there for a purpose? I don't think so. I'm an atheist.

Though they varied in what they had to say about religion and spirituality, no ninth grader explicitly rejected all religious thought. With the exception of Art who talked at some length about Native American spirituality, neither did the ninth graders refer to specific religious doctrine the way that Kelly in Cobern (1993a, p. 940) invoked a hyper fundamentalist interpretation of Genesis as a way of talking about Nature.

> **Kelly**: I believe in the Creation Theory... I think that God created everything... As you go through the Bible... Christ had to do everything that he did... he had to do this thing and he had to do this next one and it was like all in order... it needed to be done and it was done... Likewise, everything has to be done in Nature,... the big animals have to eat the little animals... little animals have to eat... the little bugs and they have to do that to survive so it... it has to be done.

It often seems to be the case that science educators assume that religious students in the science classroom will be like Kelly.[31] This study with ninth graders shows that this would be a poor assumption. On the other hand, the often presumed greater religiosity among women than among men held true.

In summary, the research did not show the distinctions between male and female students that the researchers expected to see. Regardless of gender, the students were aware of environmental problems and environmental concerns, and most students held an aesthetic view of Nature. As noted in Assertion 8, a recognizable gender factor only surfaced with respect to religion. As a group, the girls were more religious than the boys were.

---

[31] It should be noted that Kelly, although having this particular view on creation, was not found in Cobern (1993a) to be anti scientific as some might suppose.

# *Chapter 7*
## The Environment, Science and Religion

As the students spoke about Nature during the interviews, it became apparent that for most of them, the environment *is* Nature.

**Assertion 9**

As the ninth graders spoke, most of them had something to say about the environment, its status, and about conservationism.

> *The ninth grade students of this study, on the whole, articulated a high degree of environmental awareness. The relationship of that awareness with science, however, was quite mixed.*

Each of the sixteen students interviewed expressed some degree of environmental awareness in the sense that the world is experiencing significant environmental degradation and that human welfare is environmentally dependent. Students repeatedly mentioned such problems as ocean pollution, rainforest destruction, and7 the endangerment of animal species.

> **Bruce**: The natural world is exploited because of us, humans. The earth is in danger because humans are destroying the ozone layer, rain forests and precious land.... We must learn to re-use our garbage and shrink the size of landfills.... We must learn conservation techniques in order to protect our resources so to avoid the damaging effects of pollution. If we act now by not burning the rain forests and stop dumping hazardous wastes we can avoid doom and the endangerment of any more animals.

None of these topics, however, was explored in depth. For example, though students would cite the endangerment of animal species they did not elaborate at all nor even comment that plant species also suffer endangerment. Nonetheless, students showed an understanding of the seriousness of the issues. Perhaps as a reflection of the community they live in, many of the students expressed a concern for the destruction of the desert and increasing urban development that they are witness to daily. Students typically cited religious obligation or the need to protect resources as the reason for conserving Nature (see Assertion 10, page 70). While a few of them attributed science and scientists with the interest and ability to resolve some of the environmental difficulties that we face today, several students also attributed some blame to science for contributing to, if not causing, those very same problems. Many of the comments included reference to the pleasure that individual students derive from experiences in Nature. These experiences were not only family or group oriented but also ones that they had had of a more personal Nature by themselves (see Assertion 4, page 50).

Students were environmentally aware in that they recognized our human dependency on Nature. This awareness cut across the demarcations of both gender and science grade success. Consider the excerpts below, all from students with good grade success.

> **Alice**: How could someone say "Nature is just there" is beyond me because its everywhere. Nature is everywhere. It can't be something that's just there. We can't go a day without using something from Nature. We depend on Nature for everything: material items, resources, ideas and pleasures. If we as humans aren't careful, we are going to ruin the one thing that we need to survive.

> **Kevin**: Nature has always been there, but it has changed due to natural and human influences. The resources in our environment are a necessity to us for our survival. But our growing need has led to exploitation due to people's lack of caring. Overpopulation, oil spills, air pollution and ozone depletion are a result.... Our environment needs protection for the future.

> **Patricia**: Our knowledge of the natural world throughout science allows us to use our natural resources and at the same time exploit them. This exploitation will eventually put an end to Earth life as we know it, if we don't start changing our way of living. The natural world was created by God so we can serve him and care for it. We have taken advantage of it long enough.

Allen, Art, Simon, Samantha, and Betty are all students at the opposite end of the academic spectrum. They are students with lower grades and much less success in school. Nonetheless, their comments on Nature show the same environmental sensitivity.

> **Allen**: Nature gives us all we need to survive.

> **Samantha**: Nature has been exploited by many people that do not care about it. Many things have been ruined. Our earth is in trouble. It is being taken advantage of by people that are using parts of Nature that don't really belong to them. This has caused a lot of pollution and wasting of endangered species. Most of the problems that Nature has are caused by people not caring.

> **Art**: [Nature] is vulnerable to our influence and as our population has grown, Nature has been affected. Man has changed the natural world by exploiting its resources and polluting the environment. We have depleted the rainforests and changed the balance of natural things. I am concerned about the pollution we

have caused, the things we have lost from the rainforest without knowing they are there, the damage we have done to Earth's water.

> **Betty**: Nature is not going to last forever. It needs protection and conserva-tion. People should be more concerned about Nature. They don't seem to care enough about the future and what will be left for their children. The damage that people do to Nature worries me. Things like pollution and waste of resources are going to lead to more and more problems. Everything we have is made from something in Nature. People should realize that Nature is a valuable resource. As such, it will eventually be used up. The more careful we are the longer it will last.

> **Simon**: Our lack of understanding of Nature has caused us to exploit our natural resources. Ultimately we are causing permanent damage because of such things as overpopulation, oil spills, cutting down trees, pollution, etc. Possibly we are doomed.

All of these students share ideas similar to those stated by Jackie. This is interesting since Jackie is the Native American student in the group and thus the one many would expect – for cultural reasons – to have a heighten environmental conscience.

> **Jackie**: Everyone exploits the natural world. They poach endangered animals, destroy habitats, and strip the earth of our important resources. The resources are essential for life, and exploiting them not only leads to depletion, but using them leads to pollution that is destroying our ozone layer... Our natural world is endangered and people must learn to take care of it.

The students' environmental ideas were often stated without any explicit connection science. Indeed, only five students talked about science with respect to environmental and conservation issues. At that, not all comments about science were flattering. The optimistic notes came from Sally, Kevin, Samantha, and Alice.

> **Sally**: Science can teach us how to be better conservationists through research and technology so we can avoid pollution, which ruins Nature.

> **Kevin**: The resources in our environment are a necessity to us for our survival but our growing need has led to exploitation due to people's lack of caring. Over population, oil spills, air pollution and ozone depletion are a result.... I want to be a scientist. Science raises many questions about some of the changed, damaged parts of Nature. Our environment needs protection for the future. We need to protect the environment by recycling, carpooling, reducing pollution, conserving trees, etc. The ability to protect requires knowledge.... I want to be a scientist. Science raises many questions about Nature. By trying to answer those questions maybe we can learn to restore some of the changed, damaged parts of Nature.

> **Samantha**: If we all worked together we could do some restoration but I don't think that will happen.... Scientists and environmental organizations are concerned about conservation and our resources. But people just don't care!

> **Alice**: Science can work to increase our knowledge and understanding of Nature...it is all people in the world that must act responsibly to help solve the problems we've created in Nature. it will be hard for scientists to study Nature in the future if we just keep destroying it.

Patricia offers a similar perspective, but she also thoughtfully identifies several sides to the issues.

> **Patricia**: Science provides ways for us to use resources, but also ultimately exploits those resources.... science and religion have distinct roles in our life's teachings. Science teaches us how to conserve our resources, and how to possibly restore them, while religion teaches us the caring attitudes required to be productive members of the natural world.... The natural world also provides us with many resources such as, food, fuels, minerals, and plants that give us cures for diseases. Our knowledge of the natural world throughout science allows us to use our natural resources and at the same time exploit them. The exploitation will eventually put an end to life on Earth, as we know it, if we don't start changing our way of living....

Art is clearly less sanguine.

> **Art**: Nature is a source of knowledge. That's an important resource. Our ability to increase our knowledge and understanding of Nature will allow us to correct some of the damage that we have done.... Unfortunately science and scientific knowledge are also increasing our tendency to pollute, destroy and clutter up the earth and space.... Scientists also are the people that understand the need to preserve and protect.

And, in his iconoclastic ways, Art makes it clear that the knowledge of Nature he values is of a non-scientific type.

> **Art**: It is more important to have a spiritual understanding of Nature than just scientific knowledge. That understanding can't be gained from school. You have to spend time in Nature and learn to feel it. Than you will understand it.

Amongst the students there was at least one more who thought that the pursuit of science could have a deleterious effect on the environment.

> **Paula**: I don't understand the human world and why people feel the need to study Nature. Studying Nature only causes trouble. It creates more technology and curiosity, which leads to the exploitation of the land. Humans made the natural world "impure." It's our fault. I feel like humans have destroyed so much of our natural resources, that when it comes down to it, the only thing we're doing is endangering our existence. We are stripping the natural world of all its raw materials such as water, minerals, and plants vital to the Earth's

> survival. The overuse of these materials will doom us, not to mention the buildings, clothes, and machines that make the natural world "unnatural" and polluted.

Though Paula does not specifically mention science by name here or anywhere else in her narrative, the reference to "studying Nature" implicates the natural sciences.[32]

There were two other students who specifically spoke about science but not in regard to environmentalism. Their interest was access to natural resources and conservation of those resources.

> **Howard**: Scientific studies will allow us to use more of Nature to our advantage. Our exploitation has caused pollution... but I think Nature is restorable... [Nature] can be restored by natural processes if left alone. We are also able to restore it somewhat by conservation efforts.

> **Ann**: Science provides us with technology, which in turn increases our scientific knowledge. Technology helps provide us with many "wants" which, of course, increase our pleasure. It also uses resources.

Howard did not present himself as a strong supporter of environmentalism, and he was not involved with the student environmental organization called "H.O.P.E.". Indeed, he cautioned against environmental extremism.

> **Howard**: Humans have definitely influenced the natural world by building cities and communities but I think that Nature is restorable. It can be restored by natural processes if left alone. We are also able to restore it somewhat by conservation efforts. We need to be careful, though, of environmental extremism.

On the other hand, Ann was a strong supporter of environmentalism as we shall see below.

### Assertion 10

As a group, these ninth graders were not indifferent to Nature as the environment. They spoke about their involvement with environmental efforts from simple recycling to participation in a school club called "H.O.P.E.", which is a science club that does water tests for the Forest Service.

> *Most of the ninth grade students felt obliged to participate in efforts to restore and conserve Nature. This obligation was frequently expressed as a religious or personal, rather than a practical, necessity. The personal*

---

[32] Students in the study typically made no distinction between science and technology. Since the purpose of the research was to elicit student ideas, there were no attempts made to get the students to differentiate between science and technology. They were allowed to use the words as they would naturally. Of course, the misuse of the terms says something about the students misunderstanding of the nature of science, a point to be discussed in Chapter 9.

> *the obligation was primarily based on the personal pleasure derived from being out in Nature.*

As seen above, Howard was less than an enthusiastic environmentalist, but only Holly showed true indifference.

> **Holly**: People need to realize that our resources need to be protected because they are necessities for life. They can be recycled. I do not recycle because it is probably not in danger now or during my lifetime, so what's the point?

The rest of the students expressed a sincere sense of obligation to care for the environment. For seven students, this obligation was strongly grounded in the personal pleasure they derive from being out in Nature. For example,

> **Samantha**: Nature needs our conservation efforts... The pleasure I get from being in Nature is very important to me. I spend a lot of time in Nature. I'd be pretty bored if I didn't have it. It's sacred.

> **Sally**: If you truly appreciate Nature, you can see beauty in it all, which will cause you to be careful of it.

After describing the degradation he sees happening in Nature, Simon states what it is about Nature that is important to him. Like Samantha and Sally, it is personal.

> **Simon**: I really enjoy being out in Nature. It gives me good feelings. I like walking around, climbing mountains, watching a deer drink out of a river and things like that. I think about Nature and you could say I'm in touch with Nature though I understand only a little about it. I like the mystery of not understanding everything. It adds to the beauty. Nature can be peaceful, with calm breezes, lots of nice trees and no trash.

For eleven of the students the obligation they felt to care for Nature was a moral obligation – an obligation grounded in their religious convictions. Even Bruce, who appears unsure of his religious beliefs, thought it important to refer to religion in his comments on environmentalism along with his personal experiential reasons, "As humans, we have personal and religious obligations to our world to take care of it." Paula offered more about what she believed, but was still brief.

> **Paula**: God created the natural... God intended it to be here for a purpose, which is only known by Him. Because it is God's, humans have no right to mess with it.

For other students, however, the religious obligations to care for Nature were strongly felt and they said as much. Samantha and Ann both spoke passionately about the need for people to care about Nature since Nature is God's creation.

> **Samantha**: My feelings about Nature include religious feelings too. Sometimes when I think about Nature, I also think about God. These are my first thoughts when you say the word Nature. Because Nature comes from God, we have an

> obligation to take care of it.... Most of the problems that Nature has are caused by people not caring.
>
> **Ann**: To me, Nature is beautiful and pure because it is God's creation.... I also think about how caring about Nature. Because it is God's creation, we are obligated to take care of it. Nature's resources are necessities but they are also limited. It bothers me that people are so greedy and use Nature. They take things for granted and don't think about the effect that they are having on the world. Many things are now polluted and our rain forests are endangered due to lack of caring.

Patricia and Sally express similar thought. Patricia adds that it is religion's role to teach a caring attitude, and that we can serve God by caring for Nature.

> **Patricia**: God created the natural world.... Nature, or the natural world, is anything made by God, all the plants and animals on earth and the entire solar system.... Science teaches us how to conserve our resources and how to possibly restore them. Religion teaches us the caring attitudes required to be productive members of the natural world.... The natural world was created by God so we can serve Him and care for it. We have taken advantage of it long enough. People must learn to take the time to enjoy the beauty of Nature both religiously and scientifically.

Sally had the most to say on this religious perspective. While several of the other students were Christians, Sally spoke specifically about the Bible and the role that people have as stewards of creation.

> **Sally**: I think of the natural world as what God gave us to take care of. In the Bible, it says we are superior to animals and plants. So, we are supposed to take care of them. Religion teaches the caring attitude people must have in order to conserve our natural resources. We have an obligation to take care of this world because God created it for a purpose.... God intended the humans to be the superior-powerful people they are, not so they could exploit Nature but so they could become stewards of our Earth.... Sometimes people have too much power. Some of the uncaring with no religious background exploit Nature by developing nuclear bombs, destroying land, ruining our rainforests, and endangering animals which will possibly doom us. Regardless of these people, Nature will survive because of the many people who do care. People shouldn't have the power to destroy. They should only have the caring power. God would not have given all this beauty for us to ruin. We are stewards of God's land.

Finally, it is important to mention Jackie. She is the Native American student in the group and she expresses a concern for Nature that is grounded in her cultural-religious heritage.

> **Jackie**: Our natural world is endangered and people must learn how to take care of it.... The natural world is everything that God created and therefore has

> a purpose. The Indian culture believes that this land is sacred and should be taken care of.

And, as we have seen before, there is Art, the Anglo student who shows the influence of Native American culture.

> **Art**: I understand why Nature is sacred to some groups of people. It is sacred to me. The American Indian culture has the kind of understanding for Nature that encourages preservation rather than destruction.... Some scientists and Indian cultures understand their role on Earth. They do not think of themselves as superior beings and don't feel that they have a right to go around destroying Nature. They leave it the way it is because Nature was made a certain way and it is supposed to be kept that way.... Everything has a purpose that is put on this planet. We have a purpose here also. Maybe it is to preserve Nature, have fun and enjoy the spiritual quality instead of destroying it. It is our responsibility to know and under-stand human impact on the fragile, easily tampered with balance of the natural world. Nature is something felt!

**Research Implications**

A very encouraging finding from these interviews is that most of the ninth graders voluntarily expressed concerns for the environment and conservation issues. Without prompting, they self-identified these as important issues. It is also arresting that without advance knowledge one could not identify the gender of the students by their remarks about the environment. Inferences drawn from some of the feminist literature suggests that female students would be found to have a more caring attitude toward the environment. In terms of future research, this suggests two questions. Is this inference simply incorrect? Or, is the inference incorrect for these students because of other factors at work in their lives? I mentioned earlier that these students live in an environmentally conscious area. Is it possible that this cultural fact is more important than gender? The type of community in which these students live suggests further questions. If community type is a significant factor in environmental awareness, what thoughts would inner city students voluntarily expressed about Nature? What difference in thought might one contribute to economic class? What is the relationship between positive family experiences with Nature and a concerned caring attitude for the environment? If first hand experience with Nature is important for the development of environmental awareness, to what extent are school programs that bring inner city children to parks and farms helpful? Are short field trips of even a few days helpful or are the attitudes observed in this research the result of children having grown up in an environmentally aware culture?

That the students in this research expressed environmental concerns but did so with little reference to science is a mixed blessing. On the one hand, it is heartening that an environmental ethic has taken root. On the other hand, the role of science in the study of the environment is critical and therefore it disappoints that this critical association is not more evident in the students' environmental awareness. Moreover, it is positively disappointing from a science perspective to have science charged as the villain in crimes of environmental degradation. The research implications are many. To what extent do environmental studies at the elementary level make explicit the positive connection with science? To what extent does the traditional study of elementary science promote environmental awareness? What is the impact of popular media that imputes to science a pernicious character (e.g., the film *Jurassic Park*)?

Perhaps of most interest are the research implications for instruction. One begins by asking how one would want the assertions to be different. Science educators would prefer to have heard more about science. Without changing the students religious and personal motivations, we would like to hear students express their environmental awareness with more scientific depth. Though one may appreciate the ambiguous value of scientific knowledge as it is applied in the real world (the real world applications and value of science are not always clear and not always without problems and controversy), a scientifically literate understanding of the environment is an essential part of being informed about environmental issues. The question is, how does one accomplish this so that scientific concepts important to environmental issues become integrated in the everyday thinking of students? And, are we asking too much of ninth graders? In this study with environmentally aware ninth graders, there was only the faintest link between science and that awareness. At the very least, this suggests that this group of students would have benefited from elementary education that made this linkage more explicit. If the defined goal of science education is to develop a scientifically literate public that can make informed decisions on such things as environmental issues (and I think that this *should* be one of the goals for science education), then perhaps curricula need to be designed for that express purpose. Those who create science curricula should not assume that science taught at all school grade levels will automatically bring an integration of those disparate skills and concepts that makes for a scientifically literate adult. This won't happen by default. It will only occur by deliberate, carefully planned intent.

# *Chapter 8*

## The Different Worlds of Science Teachers and Their Students

There were five science teachers at the high school where the study was conducted. Three were in the life sciences and two in the physical sciences. For this study, I interviewed four of the teachers: two biology teachers and two physical science teachers. The fifth teacher was a teacher/researcher on the project, involved with the interpretation of data, and thus was not interviewed for the study. The teachers all had several or more of the 16 ninth graders of this study in a least one science course. All the teachers share the teaching of several sections of a ninth grader "introduction to science" course, and this is where all of the students in the study first met at least one of the four science teachers interviewed. The teachers individually sat for the same interview protocols used with the ninth graders. As with the students, the research wanted to know to what extent teachers enjoin scientific knowledge vis-à-vis other domains of knowledge in a discussion about Nature, given that science is unarguably relevant to the topic of Nature; and yet, Nature is a topic that most people do not explicitly associate with science? What are the concepts that appear to have scope and force in the thinking on this topic? And, how do teachers and students compare?

These questions are first addressed in assertions 11, 12, 13 and 14, which compare the teachers with their students. This comparison is important on the basis that a science teacher not only presents scientific concepts, but also creates (often tacitly) a context in which scientific concepts are presented to the class. This context can be strongly influenced by teacher beliefs and worldview. The interviews also provided insight on the differences between science teachers in different science disciplines. Assertions 14 and 15 address this insight.

### Assertion #11

In Assertion 11 we find, what commonsense says that one should find.

> *The expected is found. Science teachers show that they know more about science than do their ninth grade students. They speak more about science and they speak in more depth about science, rather than the "name dropping" speech typical amongst the students.*

| Kevin | Mr. Bradford |
|---|---|
| Nature has always been here, but it has changed due to natural and human influences. | Nature is ever-changing. If man weren't there, beavers, for example, would still change Nature! Plants invading the land have changed Nature. There are many different aspects to Nature. The more diverse Nature is the more it is the way it should be. For example, in the ideal forest, in my mind, there is a wide variety of plants and animals, all different kinds of trees. Whenever that ideal forest is altered by man, there seems to be less and less diversity among those types of organisms in Nature. Nature, however, is more powerful than the minds of people who are trying to conquer it. Eventually, Nature will win out and so far it has won out, because man has yet to conquer all of Nature and I think that... because of its complexity, because of its mysteries, because of its unsolvable Nature, it remains to be powerful, more powerful than man. Glenn Canyon Dam is an example. The dam has permanently altered an aspect of Nature. Man conquered that section of that river, but he has not conquered the entire river. Eventually the water will flow over that dam and destroy it, so Nature will be more powerful than man, in the long run. |

Figure 9

| Kevin | Mr. Bradford |
|---|---|
| I think Nature is very complex. There are unknown parts of Nature and they are confusing to me because there are no real laws controlling them. There is no order. These parts of Nature can be very powerful, dangerous and unpredictable. Earthquakes are an example, also rattlesnakes. | Due to the diversity in Nature, Nature is very complex. All of the various components of Nature are working together, and in some cases working apart. It leads toward the complexity of Nature. It makes it very hard to Figure out. It makes Nature mysterious. Nature is mysterious because it is so complex, the diversity of it makes it mysterious. There is a lot that is not known about Nature. No one will ever know everything there is to know about Nature and that is part of its appeal; because it is so mysterious. Not only will nobody ever know everything there is to know about Nature, hopefully no one ever will.... To me, the mysterious Nature of Nature is one of its better qualities. Things that are completely discovered are no longer interesting. For example, you have a cube of metal that everybody knows every single ingredient in it. Well, there is no mystery to it. There is nothing appealing about that anymore because there are no questions to ask about it. If everybody knows everything there is to know about that cube of metal, it looses appeal to me and I am sure that it looses its appeal to the people investigating it. If things don't have questions associated with them, there is no mystery. If there is nothing to ask about it anymore, it looses its intrigue, its interest. The mysteries of Nature are hopefully unsolvable. I don't want to solve all the mysteries of Nature. I hope nobody ever does. The appeal is like being lost out in the forest, so to speak. You want to be out there away from anything that is solved, you want to be in an environment where everything is still interesting to you. |

Figure 10

The full narratives for the four science teachers are given in Appendix B. Each teacher narrative is three to four pages in length. In contrast, most of the student narratives are a single page or less. In terms of brute numbers of words, it is clear that the teachers had much more to say about Nature than did their students (an indication both that they know more science and that they readily use science as a way of thinking about Nature). Consider the comparison between Kevin (an "A" student) and Mr. Bradford shown in Figure 9. Both Kevin and Mr. Bradford are talking about change in Nature. Kevin uses 15 words but Mr. Bradford uses 204 words. The difference in the number of words could be attributed to different levels of interest between the two people *except* that other examples show the same pattern. In Figure 10, for example, Kevin and Mr. Bradford are both talking about the complexity and mystery of Nature. In this second example, Kevin uses almost 50 words for this topic, but Mr. Bradford uses almost 300 words on what begins as essentially the same point.

Besides volume of words, the other point of Assertion 11 is that the teachers did not "name drop" the way that their students frequently did. Figure 11 gives excerpts from Bruce and Ms Jackson. Like Kevin, Bruce is interested in science. Though his grades are not as high as Kevin's grades, Bruce is a decent student. In Figure 11, one sees that Bruce *names* some ideas in science but does not develop those ideas in any depth. In contrast, Ms Jackson develops the idea that Nature is logical and predictable and these attributes form the basis for laws of Nature. She supports this view by giving examples using the concepts of gravity and lenses. Many of the students had even less to say about science and a scientific view of Nature. For example,

> **Holly**: the natural world is just there, you know, fish, bugs, dirt, animals, and plants.

> **Jackie**: Nature... is everything around us like plants and animals... These resources are essential for life and... using them leads to pollution that is destroying our ozone layer.

Although prompted during the interview, Jackie offered no explanation nor even any examples to go with the above assertion. She offered nothing further about plants and animals. She named no specific resources nor offered an account of how these resources are essential. Jackie offered no specific examples of pollutants though she did mention ozone destruction. In contrast, the teachers were much more thorough. In the following excerpt, for example, Ms. Jackson is sorting out the difference between teleological and functional purpose.

> **Ms Jackson**: I think things happen in Nature because of purpose... this I think is a religious view... that, not necessarily a fatalistic view, that you have no control, but that there is a destiny, that there is... as an individual, I'm contributing to it, but I'm not the biggest part of it, of human kind that there is. I think there are purposes. Animals have instincts... Humans are different than animals in that we seem to be able to reason and really take control, and again I

don't have a real biology background, but I observe those things, or watch shows on them, and so they have these cycles of their life which must have some purpose and it has a purpose for the food chain and how they are all, the whole huge, inter-related processes, so I think that when one... one animal gets killed, it's usually the weakest one, and so that they are promoting the stronger one, so that would be a purpose, and that is one of those things that happens in Nature, because of purpose.... I think that the whole inter-relatedness of us and our world, is that... like okay we do have a purpose, some bigger picture, and I think that we are playing that part, but I don't know what it is going to lead to.... I know that all things want to continue living, so that they all reproduce and that seems to be real important in Nature, for plants, people, and animals.

The pattern shown in Figures 9, 10, and 11 – and in the example of Ms Jackson – continues throughout the narratives. As noted at the start of this chapter, Assertion 11 is the *commonsense* assertion. We expect teachers to know more than their students and to be capable of greater articulation – and that is what was found in the interviews. In the next section, Assertion 12 develops a second commonsense idea.

| **Bruce** | **Ms Jackson** |
|---|---|
| Chemists and astrophysicists can come up with theories based on the order and the predictableness of the phenomena in the world. Some things scientists know about are: weather patterns, El Nino, Ozone depletion, and tectonic plate movements. | I think that Nature is predictable. I think that it is logical. I think that it is explainable. As scientists, we come up with laws of Nature or theories of Nature to be able to predict behaviors and therefore, based on what we know, and the experiments that we have done... there are certain patterns that we can find, and yet at times they can be very complex. But I think Nature, you can understand it... gravity is, and you throw a ball, then you can predict what is going to happen because it is logical.... If you were going to use lenses, and you know how light is going to go into the lenses, then let's say, with a telescope, you know that you need to use a lens to get the image and then the lens to invert the image, and so how you know how to use it, because of what it does. It is very logical to me. |

Figure 11

**Assertion #12**

A principle question of this research is about the extent to which students and science teachers voluntarily enjoin scientific ideas (vis-à-vis other types of ideas) in a conversation only tacitly related to science. Assertion 12 speaks to this question.

> *When compared with their students, science teachers have a much more focused and less diverse conceptualization of the natural world. As a proportion of their total comments, students speak much more about Nature as the environment, and much more of the aesthetic aspects of Nature and spiritual or religious ideas associated with Nature. In contrast, their teachers are more focused on what one can know about Nature through science.*

As noted in Assertion 1 (page 38), the ninth grade students tended to discuss the natural world using several different perspectives such as religious, aesthetic, scientific, and conservationist perspectives. A breadth of perspectives typically characterized student discussion about Nature. One might say that the students know a lot about many things pertaining to Nature but nothing in depth. Patricia is good example (Figure 12).

| Patricia | Mr. Hess |
|---|---|
| God created the natural world. It has many characteristics: it's powerful, diverse, changeable, and beautiful (physically and emotionally). The Bible says God created the heavens and Earth, so I think that explains to me what Nature is.... Even though Nature is mysterious, everything is knowable but maybe not for us now, or even in the near future. The wonderment of the world increases knowledge through science, but is limited, due to its complexity.... | Nature is orderly and understandable. The tides and the rotation of the earth... That the planets and the stars are governed by physical forces and any deviations are simply because we have not yet discovered the other part of Nature's orderliness... As a science teacher, I feel that with enough scientific knowledge all things are understandable... I think that the more we understand about matter itself, and the more we know about how to make things, the more predictable Nature will be. Scientific or reductionistic thinking is very powerful. I feel that once we know enough about the minutia of the world, breaking it down by using the scientific method, scientists tearing it apart and analyzing the parts of Nature and seeing how they interact, that we will be able to predict just about anything about Nature. |

Figure 12

Glancing at only the beginning of her narrative one quickly sees religious and aesthetic elements in addition to science elements. In Figure 12, one also sees the contrasting opening comments from Mr. Hess. When the interviews began, the science teachers almost immediately lapsed into science talk. This is a critical observation given that the interviewer is never the first one to raise the issue of science during an interview. Science only enters the interview discussion when the informant raises it. For all four of the science teachers, the topic of Nature immediately brought science to mind. Figure 13 shows a second example of this phenomenon. Mr. Bradford is a biology teacher and later we will see the marked contrast between biology teachers and physical science teachers. But here, we see that all of the teachers are alike. They all very quickly lapse into science talk about Nature, which is in marked contrast to most of the students.

The broad outlines of this assertion are not surprising. These are teachers of high school ninth graders and one would expect them to have more to say (teacher narratives run 2 to 4 times the length of student narratives) than their students and have more to say about science. It is nonetheless interesting to note how focused the teachers are on science given that no direct science questions are asked during the interview. The contrast with their students is unmistakable. This raises the question of whether the contrast between teachers and students is of any instructional significance? Does a teacher's outlook on Nature influence his or her behavior in the

| Alice | Mr. Bradford |
|---|---|
| How could someone say "Nature is just there" is beyond me because its everywhere. Nature is everywhere. It can't be something that's just there. We can't go a day without using something from Nature. We depend on Nature for everything: material items, resources, ideas and pleasures. If we as humans aren't careful we are going to ruin the one thing that we need to survive. I see many sides to Nature. It is material...full of resources: It is living...it can be hurt. I think Nature is very dangerous. I think Nature is very beautiful. It can be beautiful and peaceful but also dangerous and frightening. | Nature is the living and non living components of the world around us - even the universe apart from the works of man but including man. Nature has been here forever and it will always be here whether man is here or not. We are all interconnected in Nature by both natural and artificial mechanisms. We, as a species, are all part of the same natural world and we all have an effect on each of the other components and species of the natural world. Nature in its natural state is pure and perfect. It can improve itself but it cannot be improved by the works of man. Purpose in Nature means the struggle for survival. It does not mean there is a god who gives purpose to Nature and directs Nature by will. |

Figure 13

science class in such a way as to also influence students who may have very different conceptualizations of Nature? In Chapter 10, I will offer a classroom example that suggests an answer. Despite this difference, with respect to science between science teacher and student, there are similarities between teachers and students and these are addressed in Assertion 13.

**Assertion #13**

As noted above, aesthetic and religious ideas contribute a greater proportion to the student narratives than to the teacher narratives. Nevertheless, these elements were present in the teacher narratives.

> *Most of the teachers spoke about religious or spiritual ideas with respect to Nature, but the teachers were typically more agnostic and more philosophical than their students.*

The four science teachers fall into three categories with respect of their comments on religion and spirituality. In one category, there are the two teachers who affirm, in relatively strong terms, a connection or association between religion and Nature. In a second category, there is the one teacher who essentially equated spirituality with aesthetics and mystery. In the third category, there is the indifferent teacher.

None of the teachers declared that God made the world or that God created Nature as did some of the students. Still, two of the teachers spoke at some length about religious ideas and Nature that were important to them. One can see this in the excerpt below. In this excerpt, Ms Jackson also connects religious thought and environmentalism.

> **Ms Jackson**: Nature brings out emotions and I think that because of that we should respect it, definitely from a religious side as well. And I think that, you know, and that goes for our own bodies... everything. That's because that's more of a religious side that we should take care of every-thing. I think that we should protect nature... you know there is that saying that we are just borrowing the earth from our children, they're not giving it to us.... I think that the whole inter-relatedness of us and our world, is that... like okay we do have a purpose, some bigger picture, and I think that we are playing that part, but I don't know what it is going to lead to.

The second teacher in this first category is Mr. Hess. That Mr. Hess should be in this category is of considerable interest. Mr. Hess is the teacher who spoke most about logical, empirical, and reductionist thinking, but he adds to his comments on Nature a different perspective.

> **Mr. Hess**: I also have some other thoughts about nature that are really completely separate from what I have said so far. These thoughts are extremely important because I think that there is a need in man's life for a purpose. Nature or the natural world is everything. Well, it's the universe, including man, and everything man does, and everything in the universe; but, the natural

world is not everything that exists. I think God exists and He is part of the natural world, but at the same time, not part of it. I think that the natural world is a subset of God, and not the other way around. I think that nature can remind us of the spirituality, our own spirituality.

He is careful not to go too far in this direction of spirituality for he says:

**Mr. Hess**: I don't think that nature has a spiritual quality in itself. I don't think the world around us or the universe has any spiritual qualities.

Mr. Hess is not a New Age enthusiast and would likely give little heed to eco-spiritual ideas such as the Gaia hypothesis. He is reminiscent of Isaac Newton in that his perfect system of Nature has a requirement that is fundamentally *other* than material Nature.

**Mr. Hess**: So, God is sort of the wrench in that perfect definition, nature or the natural world is everything, but a necessary part. It is a necessary wrench because the rest does not work without that.... Divorced from pure science and pure fact there is also something called faith which is what defines sacred and holy and mysterious.... I definitely think that there are parts of everything that are separate from, not Nature, or the natural world, but definitely from what I perceive as what science can uncover, and maybe part of that has to do with man's need and wanting for, and maybe personal discovery of things that are holy and sacred, or mysterious. That is sort of an unknown variable, that sort of sits out there. These aren't products of man's interaction with that part of the natural world. Things become holy because they are a part of the natural world that doesn't fit into a nice little formula, but is somewhere out there that we haven't tapped into. Maybe that is knowable, maybe that is predictable, eventually, but it can't be broken down by using the scientific method.

So here is Mr. Hess, the physicist and the teacher who most closely matches the stereotype for science teachers, and yet he is the most traditionally religious of the four teachers. It is also interesting to note that both teachers in this first category – the most religious of the four – are the two physical science teachers, Mr. Hess and Ms Jackson.

Mr. David is the other teacher who spoke at some length about spiritual ideas but in a way very different from Mr. Hess or Ms Jackson. Mr. David is in the second category. His religious ideas about Nature are essentially aesthetic and emotional ideas.

**Mr. David**: Some people might say they see the work of God in Nature... to appreciate that is one of the aesthetic things that we like about Nature. I have an instinctual connection to sacredness of Nature. It just deals with something very special about Nature, and you have to respect Nature. As I said before, I think that it is important to understand that there is more to Nature than meets the eye. We need to treat it as being sacred, because of that. I think that Nature is our home and it gives us life, and everything, that people consider it to be very sacred and holy, and those are definitely the perceptions that we have as being a

> part of Nature, and it kind of grows out of that, I think. These ideas are religious and philosophical and emotional, all three.... I think that people have real strong emotional ties to Nature, in a lot of ways. There is a lot of variations on how people consider it to be sacred or holy, but I think everybody does, in some sense.... Everybody has a sort of innate appreciation for life and for the natural world. I think that a lot of our religious belief has to come from this aspect of Nature, its beauty and mystery and sacredness, which is that part that we can't... or we don't understand at this point, and that it makes us think that there must be something beyond our level of understanding, from where Nature must have come. You wonder about how it all got started and where it all came from? Its very religious, emotional, and philosophical too.

After the interview with Mr. David, who is a biology teacher, he commented that where he used the word "God" he really meant "beauty." For example, where some people see God in Nature, Mr. David sees beauty in Nature. He was quite agnostic about any traditional theistic concepts of religion.

> **Mr. David**: I look at the natural world as something that is physical... but if there is a God behind it, that is creating it, then that might be something that is at a different level, that I don't understand in my own reach [pause] from what we know about energy and physics and everything, I think that there is potentially other realities or perceptions, or things that are happening, that are beyond this natural world that we are perceiving. I think that there could be things that I just don't know about. I do think Nature is more than material.

The natural world is physical and he seems to see this as being at odds with the idea of a creator God yet he holds open the possibility. Mr. David finds room for that possibility through his understanding of physics. In the end when he asserts that Nature is more than material, one suspects that he has the aesthetic in mind. Thus here we have a biologist – in contrast to Mr. Hess the physicist – whose religious ideas about Nature are strongly influenced by his science and his sense of the aesthetic.

The last category is the category of indifference. Of the four teachers, Mr. Bradford, also a biologist, clearly had the least to say about religion and spirituality. Indeed, he never mentioned God. He never talked about spirituality. When he used the word "purpose" it was about functional purpose. His only comment about religion was essentially a denial of any transcendent purpose in Nature.

> **Mr. Bradford**: Purpose in nature means the struggle for survival. It does not mean there is a god who gives purpose to Nature and directs Nature by will.

In summary, with respect to religious ideas one sees a diversity of thought among the four science teachers and diversity with respect to their students. On the whole, the teachers appear to be less religious in traditional ways than their students. The physical science teachers, however, are the more traditionally religious of the four teachers. In Assertion 14, we will see again differences among the science teachers but we will also see strong similarities with their students.

**Assertion #14**

I have noted the strong environmental and aesthetic views among many of the ninth grade students. The teachers held similar views though there are intriguing differences between the physical science and biology teachers.

> *The science teachers like most of their students had a strong aesthetic sense of Nature. They also had a strong sense of Nature as our environment. The biology teachers, however, had a more conservationist view of the environment while the physical science teachers had a more resource view of the environment.*

Under Assertion 6, we saw the ninth graders' strong aesthetic sense of Nature. On this point, the students and their teachers agreed substantially. Nature is beautiful. Nature is an aesthetic experience.

> **Mr. Bradford**: Nature is beautiful, as I see it. That is what draws me to Nature in the first place, how beautiful it is. The simple beauty of being pure, the kind of plants and animals that are out there, the landscape in its natural state, all kinds of simple beauty to it. Purity and diversity have an internalized beauty to me. When it is pure and when there is great diversity out there, then it is more beautiful to me. So, those things have to be in place first, possibly before I consider it to be beautiful. Nature is living. Nature is composed of living things organisms, and the living part of Nature is probably what attracts me to Nature in the first place. So, the living part of Nature is what appeals to me plants and animals, any kind of plants or animals. Even though I would consider rocks and volcano's a part of Nature, the living part of it appeals to me more. There are some beautiful rock formations and so on, but the living parts of the landscape is what is most beautiful. The appeal for me is an internal sense of peacefulness when I am around Nature.

> **Mr. David**: I like the word beautiful. I think that there is a lot of beauty in Nature, even though it is not always beautiful to man. The whole aspect of Nature and I guess that I have an instinctual connection to that, that it is sacred, and just deals with something very special, you have to respect it. I think that beauty is the more aesthetic reason to appreciate Nature and I think that aesthetics can provide reasons for studying Nature, too. But, I think that beauty and emotional response are more in the aesthetic realm, just pleasing to see how Nature works, seeing that it is mysterious, that it allows a curiosity about how it works - to admire the beauty of Nature and it's simplicity, and just... Well, I enjoy Nature.... it is so amazing and interesting, to see and to experiment, that any curiosity that we have about it is a good enough reason to go and study it. There are aesthetic reasons. It is just pleasing to see how Nature works.

If there is one salient difference between student and teacher, it is that the teachers were more likely to connect aesthetics with science. This can be seen in Mr.

David's excerpt above and more plainly in comments from Ms Jackson and Mr. Hess.

> **Ms Jackson**: I've lived in a variety of places. I have lived around a lot of mountains, around the Alps and I've lived in the Blue Ridge and Appalachian mountains of Virginia, and it has taken me a while to appreciate the beauty of Arizona... just in a natural setting, and I think Nature is beautiful. I think about Nature everyday in one way or another. If it's not the laws of Nature, driving with my kids and I am pointing out the moon to them in Arizona, and like I've said that I have lived a lot of different places, and the sunsets here are the most beautiful sunsets, and I know why we see those sunsets, but it is just nice to enjoy them. I also think that science is beautiful in the fact that you can repeat patterns and that you can find these things that are logical and I just like that. That appeals to me. Because of the physics and the refraction of light you can understand a beautiful sunset.

> **Mr. Hess**: Nature is beautiful. I see it most in the way things work so well together. I think that I see beauty in Nature more with living things than with anything else. It is the vastness of things that could go wrong in a living organism, and yet it lives.

Thus, on the idea of aesthetics in Nature the teachers are largely in agreement. With Nature as the environment, one begins to notice differences.

Like the ninth grade students, the science teachers – as one would expect – all spoke about the fragility of the environment. The following excerpt comments are from Mr. Bradford and Mr. David, the biology teachers. In the long excerpt from Mr. Bradford's narrative, there is an acute sense of concern for the environ-ment and it is notable that in this context he does not speak of Nature as a resource for humans.

> **Mr. Bradford**: we are effecting Nature all the time. For example, we effect one thing and it effects several other things that will have an effect on me or other parts of Nature. This indirectly affects me again. It is important to me to have people understand how Nature works so that they can prevent effecting it more than they do. Nature is difficult to understand remember, it is very diverse. People think that they understand Nature and so they go ahead and affect it the way that they want to. They predict that they won't affect anything else, but in fact, they do. So, I think that it is important to understand how Nature works, as best we can, so as not to destroy it. I think everybody should study Nature, I am not sure that everybody does do it. Scientists probably do more of it than anybody else. Nature, however, is really not knowable. If Nature were knowable, it would mean that you would be able to predict anything about Nature. You would know all the components of Nature, what effects it has on other components. Well, you can predict a certain amount of Natures' outcome, but overall, Nature is unpredictable. If we destroy this insect, for example, because it is hurting some tree or some plant that we care about, I don't think that we will

ever be able to predict how we have effected Nature as a whole. You can't predict what effect one aspect of Nature will have on all of the other components of Nature. So it is unpredictable in that regard. Because man has shown that he can change certain components of Nature, I am concerned about pollution and the damage that it does to Nature. Right now there is overwhelming damage being done to Nature - the effects of man on Nature in our lifetime are pollution, destruction of rainforest, the damming of rivers, the dredging of the oceans, and the pollution of our oceans, the killing of species, and so on.

The comments from Mr. David are similar.

> **Mr. David**: Because of our dependence on Nature, just our existence, we need to study Nature, to learn more about it. We need to understand how things work in Nature because it is an important resource for us, to get our water, energy, food, and materials for making things from Nature. The resources that Nature contains is kind of unpredictable, because we don't really know what resources are there, that we can use. Because of that, we need to understand it as much as we can, so that we can protect it. It needs to be protected... and keep it so that it can maintain us and maintain civilization.

Unlike Mr. Bradford, Mr. David speaks about needed resources; but he does not speak at length on resource use and he is very quick to point out the danger of exploitation.

> **Mr. David**: Man has an impact on the natural world. Because the world is full of resources and powerful, man has also polluted and exploited it, even though it is powerful, and has taken the resources and used them for his own purposes and things. Although I think everybody has a sort of innate appreciation for life and for the natural world, when people have self-interest at stake, they tend to meet their own needs.

On the balance, the two biology teachers appear to be much more concerned about environmental issues and potential damage to the environment than about resource issues.

The physical science teacher, Ms Jackson, shows some of the same environmental concerns but the concerns are voiced in the context of resource use.

> **Ms Jackson**: I am hoping that there is a balance between use of resources and protection just because this awareness of Nature keeps being raised, and people keep coming up with solutions.... I think that we should protect Nature... you know there is that saying that we are just borrowing the earth from our children, they're not giving it to us.

Moreover, these are Ms Jackson's only conservationist comments. Her other comments on Nature as the environment are about the environment that provides natural resources. The other physical science teacher, Mr. Hess, holds a similar view. Nature is about resources; Mr. Hess actually uses the word "exploit" favorably.

> **Mr. Hess**: Nature is made of matter. That matter gives us the resources we need whether it is living resources or material resources. Material resour-ces are the raw materials that we can use to build things or to develop technology. Thus, the second reason to study Nature is that the more we know about Nature, the more we can control it and use it or exploit it.

The protection of Nature is solely for the protection of human interests.

> **Mr. Hess**: We need to protect Nature so that nature can provide us with the materials we need. So, if you put man into the equation, like if the equation says that man needs to be on this planet, then this is what we need to do. If we are not concerned with that, then we shouldn't really worry about what we do with Nature.

Mr. Hess goes on to say that the protection of Nature is a kind of balancing act.

> **Mr. Hess**: I think there is a bigger story, though, about why we need to protect and know about Nature. This is such a bigoted statement, but we need to protect the human race. We need to know enough about the eco-systems, so that we can say, "yes, these animals can become extinct because they are not really important." So, those two sides of myself battle each other because I think that there is a lot of beauty in Nature and I think that it is very enjoyable to have these diverse animals. But, I also think that we also need to be realistic and know that we are not going to be able to protect all of the animals. We need to know what animals are necessary for us to enjoy the same quality of life that we now know.

In summary, each of the four science teachers spoke about the aesthetics of Nature and each voiced some environmental concerns. In these respects, the teachers are like their students. Overall, the two biology teachers appeared to have a stronger aesthetic sense of Nature than the physical science teachers, and were much more concerned about environmental issues.

**Assertion #15**

Assertion 14 addresses some of the differences observed between the physical science teachers and biology teachers. In contrast to the biology teachers, the physical science teachers had distinctively different conceptualizations of Nature. The two physical science teachers were more religious, but the two biology teachers had a more acute sense of environmentalism. The most striking difference, however, was an epistemological difference.

*Physical science teachers spoke much more about all that scientists* ***do*** *know about Nature and how successful science has been. Biology teachers were* *also enthusiastic about science. The biology teachers, however, were much less sanguine about science.*

The two physical science teachers of the study show a strong sense of order and logic in Nature.

> **Mr. Hess:** Nature is orderly and understandable. The tides and the rotation of the earth, the seasons and so forth are examples of order in Nature. That the planets and the stars are governed by physical forces and any deviations are simply because we have not yet discovered the other part of Nature's orderliness. According to chaos theory, even things that appear to happen randomly have patterns. I think that everything has patterns.
>
> **Ms. Jackson**: I think that Nature is unpredictable. I think that it is logical. I think that it is explainable.... we can predict those things because they are orderly, there are certain patterns that we can find, and yet at times they can be very complex.

As one might expect after hearing such strong affirmations of order in Nature, these two teachers are also highly confident about our human capacity to gain knowledge about Nature through science. Mr. Hess' remarks are continued in the following excerpt. Note the confidence he shows in scientific knowledge.

> **Mr. Hess**: We haven't necessarily discovered those patterns, yet. As a science teacher, I feel that with enough scientific knowledge we all things are understandable.... Scientific or reductionistic thinking is very powerful. I feel that once we know enough about the minutia of the world, breaking it down by using the scientific method, scientists tearing it apart and analyzing the parts of Nature and seeing how they interact, that we will be able to predict just about anything about Nature.... I think there is probably a limit to predictability in Nature. I think Nature has unpredictability because it is so changeable.... I think unpredictability, however, comes because we don't know enough about Nature to predict everything about it right now.... Eventually, however, all Nature will be explain-able.... One of the reasons why we don't yet understand enough about Nature, is because the extreme, complex, and diverse type of systems that are involved with it. But I am an optimist as far as it's understandability, as far as that is concerned. Our current state of being is that there are unpredictable events in Nature. Our ultimate state, the end point, is basically knowing very much. Weight wise, we are probably more tilted toward unpredictableness because I think that we are in the infancy of understanding the world around us. I am optimistic that we will eventually know much more. As knowledge grows we will change the changeability and the unpredictability of Nature. It all will decrease significantly. I have a great faith in man's ability to understand things and take things apart, to get to the bottom of the solutions and things. I think that with that know-ledge and the yearning for knowledge, whatever is the problem, we will basically be able to know and being to predict.

Ms. Jackson is equally confident.

> **Ms Jackson**: As scientists, we come up with laws of Nature or theories of Nature to be able to predict behaviors and therefore, based on what we know, and the experiments that we have done, we can now, either change or know that we can't change an event, but that maybe we can predict that the event is going to happen.... As scientists, we come up with laws of Nature or theories of Nature to be able to predict behaviors and therefore, based on what we know, and the experiments that we have done, we can now, either change or know that we can't change an event, but that maybe we can predict that the event is going to happen. And we can predict those things because they are orderly, there are certain patterns that we can find, and yet at times they can be very complex. But I think Nature, you can understand it, you can know it, and you can predict it. I think that if we study it that Nature is not difficult to understand. For instance, I am not a real biology type person, but I like watching those shows and they show patterns of things having these five sides, so I guess that if you are to get a new plant, then you could categorize it, because of those sides, but basically, from the physical science side, a lot of, if you are going to... gravity is, and you throw a ball, then you can predict what is going to happen because it is logical... I think that is what I'm thinking of when I think orderly... I think logical. If you were going to use lenses, and you know how light is going to go into the lenses, then let's say, with a telescope, you know that you need to use a lens to get the image and then the lens to invert the image, and so how you know how to use it, because of what it does. It is very logical to me.... I feel like we know an awful lot. I feel like, that, even though we don't have all the answers, we have so much, ways of finding out answers. I think we have that base of knowledge, so I think, I feel that we know an awful lot. I think we would definitely be up towards the 80 percentile of knowledge.

It is very interesting to note that both of these teachers use the phrase "taking things apart," that is, you can learn about Nature by taking it apart. Mr. Hess specifically refers to a reductionist view of Nature. These are the things these teachers say about Nature. On the other hand, as noted under Assertion 14 above, Mr. Hess had much less to say about the aesthetics and emotional aspects of Nature. Ms. Jackson has somewhat more to say about the aesthetics of Nature, however, she does not venture far from intellectual beauty.

> **Ms Jackson**: I also think that science is beautiful in the fact that you can repeat patterns and that you can find these things that are logical and I just like that. That appeals to me. Because of the physics and the refraction of light you can understand a beautiful sunset.

In summary, the physical scientists view Nature as logical and orderly. Some might say is to be expected since physical scientists study inanimate objects, which are inherently more predictable than living things. The physical scientists are highly confident that science has told us much about Nature and eventually will tell us very much more, if not all. They have a muted sense of the beauty of Nature; moreover that sense of beauty is primarily of an intellectual type. The biologists could hardly be more different.

To begin with, the world of a biologist seems a much more complex place than the world of physicists. Note Mr. Bradford's use of the word "mysterious."

> **Mr. Bradford**: Due to the diversity in Nature, Nature is very complex. All of the various components of Nature are working together, and in some cases working apart. It leads toward the complexity of Nature. It makes it very hard to Figure out. It makes Nature mysterious. Nature is mysterious because it is so complex, the diversity of it makes it mysterious. There is a lot that is not known about Nature. No one will ever know everything there is to know about Nature and that is part of its appeal; because it is so mysterious.

Mr. David also uses the term mysterious. Moreover, he invokes the Eastern cultural concept of Ying/Yang to describe the attributes of Nature.

> **Mr. David**: Nature has many aspects. It is alive and it is always changing. It has a mind of it's own and in some ways things happen, because it is alive. Just the way that the earth moves and shakes, the way that the oceans tend to move and the whole relationship between the earth and the universe. The way that living things have come out of all that, or part of it, to interact with the earth and universe. I think that the fact that it's alive really is a big part of what makes it the natural world, or at least my concept of it. I am not using "alive" in the technical living things sense, but I think in terms of how matter (Nature is material as well) interacts. I think that it is alive in the sense that, even though it may not technically be alive, I think that when there is heat and there is energy, things are moving and flying - that in a way is a kind of life. Nature is dynamic... movement and change and all life, when you look down to the molecular level, it really is just non-living, material molecules that are organized in complex ways. So, it is hard to draw the line, when you get to that level, as far as what is alive and what isn't. So, that's partly what makes it mysterious. Nature is alive and it is material.... Nature is orderly and chaotic, predictable and unpredictable - these pairs are sort of needed in order to define each other. Things wouldn't be predictable if you didn't know what unpredictable was. Things wouldn't be orderly if you didn't know what chaotic was. It is sort of a Ying/Yang relationship between the two – I would call this just the dualistic Nature of reality. A storm in the ocean might be considered chaotic, but then as you watch the ripples of the waves that are flowing away from it, there is a sort of orderliness to that. Weather is unpredictable. You can't predict what's going to happen, but you can predict the consequences of it.

Just as the physics teachers' confidence in the innate orderliness of Nature led them to a highly confident view about knowledge of Nature, the biologists' view of the complexity of Nature leads them to a muted confidence in what can be known about

Nature. Mr. David, for example, sees new discoveries as inadvertently leading to even greater complexity.

> **Mr. David**: There is a lot of diversity and complexity in Nature, and there is also the fact that it is just there. It's all just part of everything that is there. You can look at it all as being part of one thing, or you can look at it all as being different and complex in different aspects of it. It is incredibly complicated. The closer you look the more complicated it is and in order for it to function as simply as it appears to us, there must be a lot more to it than we know. I think that it is important to understand that there is more to Nature than meets the eye. It is interesting to see how Nature works and just how complicated it really is. By observation and by looking at things and watching them over a period of time, you begin to notice patterns that allow you to make predictions. But it seems like a lot of predictions, once you make them you find that they... well, the rules tend to get broken, or you get more information at a higher, finer, more detailed level and you realize that there are other things going on that you weren't predicting.

Mr. Bradford makes the interesting comment that he actually prefers that some things in Nature remain unsolved.

> **Mr. Bradford**: Not only will nobody ever know everything there is to know about Nature, hopefully no one ever will.... To me, the mysterious Nature of Nature is one of its better qualities. Things that are completely discovered are no longer interesting. For example, you have a cube of metal that every-body knows every single ingredient in it. Well, there is no mystery to it. There is nothing appealing about that anymore because there are no questions to ask about it. If everybody knows everything there is to know about that cube of metal, it looses appeal to me and I am sure that it looses its appeal to the people investigating it. If things don't have questions associated with them, there is no mystery. If there is nothing to ask about it anymore, it looses its intrigue, its interest. The mysteries of Nature are hopefully unsolvable. I don't want to solve all the mysteries of Nature. I hope nobody ever does. The appeal is like being lost out in the forest, so to speak. You want to be out there away from anything that is solved, you want to be in an environment where everything is still interesting to you.

Mr. Bradford's reference to the appeal of what is unknown in Nature leads to another difference between the biologists and physicists. The appeal of Nature mentioned in the above excerpt is similar to the intellectual appeal sensed by Mr. Hess and Ms. Jackson, though Mr. Bradford uses a word, mysterious, that was used by neither Mr. Hess nor Ms. Jackson. Again, as noted under Assertion 14, it is the two biology teachers, in contrast to the physical science teachers, who have the stronger aesthetic and emotional understanding of Nature.

> **Mr. Bradford**: Nature is beautiful, as I see it. That is what draws me to Nature in the first place, how beautiful it is.... I like the word beautiful.

In summary, the two biology teachers appreciate the scientific study of Nature, they would even say it is appealing, but they also have a much more limited confidence in what science can tell one about the natural world. The power of science notwithstanding, the natural world remains a mysterious place, and it should be so says Mr. Bradford. Mystery is part of Nature's great appeal. That Nature is *appealing* is another important aspect of how the two biology teachers conceptualize Nature. Nature for the two biology teachers is a place of considerable aesthetic beauty and peace. In this sense, they are more like their students than are the two physical science teachers.

### The Nature of Science

As one can easily see in the Assertions of this chapter and the narratives in Appendix B, as the science teachers talked about Nature they provided glimpses into their thoughts about the nature of science. Indeed, there are interesting observations to be made about what the four teachers understand the nature of science to be. The teachers are a well educated cadre of science teachers. They are known in their local education community as good science teachers. Yet, clearly they do not hold in common a single homogeneous view of the nature of science (Cobern & Loving, 2000b). Is one or two of these teachers more correct in his or thinking than the others? If the science community subscribes to a single threshold view on what minimally counts as an orthodox interpretation of the nature of science, as suggested by Smith and Scharman (1999), then these four teachers likely quality for orthodoxy in spite of their differences. But I am more impressed by their differences and am inclined to take their ideas as evidence for Alters (1997) thesis that the nature of science depends on who you ask to define it, even when those you ask are qualified science teachers. In Chapter 9, I will suggest that rather than pursuing orthodoxy, it would be a better teaching strategy for science teachers to explore with their students the many different ways that science is interpreted, and include in that exercise the exploration of reasonable limits to acceptable diversity.

# Part III: Putting Things Together

# *Chapter 9*
## Limitations & Implications for Research

As noted at the book's beginning, this research takes it as axiomatic that the more educators know about "students as people," the better educators will be able to teach "people as students" in their classrooms. In order to know students as people students must be allowed to speak for themselves. This research constructed a situation where students and their science teachers could freely express their thoughts and ideas about Nature. The previous pages are filled with those ideas as well as the lessons (in the form of assertions) drawn from what the students and teachers had to say. The research, however, raises a number of additional questions, questions that also serve to set the limitations of the research I have reported in this book. Viewed as limitations, it is appropriated to discuss these further research issues prior to the general discussion in Chapter 10 that brings together the various ideas and findings of the research. In the following paragraphs, I have identified nine areas of research implied by the present research project. This list is substantial, though surely not exhaustive.

**(1)** Since the students in this research were ninth graders, we must ask whether student conceptualizations change with increasing age and maturity regardless of school experiences. In other words, were the thoughts that students expressed in these interviews largely a function of the immaturity of ninth graders? If one were to interview these students as twelfth graders, would there be grounds for the same set of assertions? Previous research with college students suggests that many of the assertions would likely remain the same (see Cobern, 1993a).

**(2)** At the end of Chapter 5, I rhetorically asked, "So, what was it about their school science experience through eight grades that science should have such a minimal impact on their *everyday* thinking?" There is much to be said in favor of the concept of scientific literacy, but as my rhetorical question suggests, scientific literacy cannot be evaluated solely on the basis of standard science achievement exams. Science education research must look at the interplay of ideas and the impact of science on everyday thinking. Simply improving standardized test scores will never be enough.

**(3)** A fascinating aspect of this study is the glimpse one is afforded into the teachers' thinking about science. The literature convincingly argues that there are many less skilled science teachers who have views on the nature of science that are inadequate (Lederman, 1992 & 1995); but as noted in Chapter 8, the four teachers of this study show that even qualified, skilled science teachers can have different ideas about the nature of science. Hence, one must ask how much nature of science orthodoxy should be expected of a science teacher? Smith and Scharman (1999) answer with a minimalist view on the nature of science for which they claim "general agreement among science education stakeholders" (p. 493).[33] Their view, however, greatly under-emphasizes the human element in science. They speak of science as if it could exist without scientists, science teachers, or science writers; or as if science can always be unambiguously disentangled from the morals, ideologies, and prejudices of scientists, science teachers, and science writers – in other words, as if science could exist in a cultural vacuum. It is a principal assertion of this book (and challenge to researchers) that the science education community should strive to better understand science from a cultural point of view and that neither student nor teacher cultural perspectives should be masked in the classroom, but made part of the instructional process.

**(4)** Similarly, the study offers a glimpse into the students' thinking about the nature of science. Not surprisingly, there is a range of ideas. Some students hold both scientific and religious ideas without discomfort. For at least one student, science can disprove religion. For some students, science is an environmental ally; for others science is an environmental enemy. At least one student seemed to be a young positivist, so confident was he about the power of science. None of the students distinguished between science and technology. Here is a group of students with perhaps a bit too much diversity of thought! Some of their ideas are valid. Other ideas are quite misinformed. The question for research is how does one teach the nature of science such that students become informed about the standard practices of science and scientists, while at the same time allowing students to explore the kind of diversity evident in the teachers of this study, among other teachers, and among scientists?

**(5)** What should one expect from a science curriculum? How should scientific literacy be defined? What type of science teacher provides the best model, a Mr. Bradford or a Mr. Hess? In the study, Howard is most like the "potential scientist" if one has in mind Mr. Hess. But are the ideas about the natural world held by Howard and Mr. Hess a good example of what one wants to see in the public? What about Mr. Bradford's view of the natural world? I raise this as a policy question concerning how science should be understood and how scientific understanding should be integrated with the various ways there are of knowing about the world.

---

[33] I do not disagree with this claim. I suspect that their view is widely shared, but I do not think this is necessarily a good thing.

(6) Chapter 8 discusses the critical differences that were observed between the students and science teachers in this study. Both Allen (1995) and Larson (1995) show similar differences between teacher and student perspectives. However, only four teachers were interviewed in the present study. To what extent are they representative of other science teachers? Clearly, more teachers should be interviewed. Moreover, the significant differences observed between physical science and biology teachers in this study was at least as interesting as the differences between students and teachers. Hence, are the differences I observed between teachers in these subject areas a common characteristic of physical science and biology teachers? If these characteristics are commonly found, do they also influence the classroom environment and the differential response many students have to the physical sciences in contrast to biology?

(7) Do these characteristics of students and teachers matter in the classroom? In the science classroom, what interactions can be better understood by explaining the interactions in terms of teacher/student conceptualizations of Nature, if indeed any? The suggestion here is that conflicts between student and teacher conceptualizations of Nature will have an adverse affect on the student's experience in the science classroom. For example, one might ask how well does an aesthetically minded person fare in Mr. Hess' science class? Is this another situation where the metaphor of *cultural border crossing* would be useful for understanding the interactions among teachers an students? In the next chapter, I will take up this very suggestion using an example from the study.

(8) There are also questions that arise given the demographics of the school and community in which the research was conducted. What would student conceptualizations of Nature be like in a very different community, for example, in an urban low income community? Would the same set of assertions be obtained? There are special programs designed to give urban children personal experiences with the rocks, trees, and forests so foreign to big city life (e.g., The New York Times, 1995). Do these programs improve interest in science? They may not if school science is perceived to be so abstract that it is no longer recognizable as the study of Nature. I noted under Assertion 4 the importance of students' personal experiences in Nature. Thus, one may infer from this that science educators would do well to make a more explicit connection between school science and student experiences as suggested in (say) Cobern et al. (1995b) or Grumbling et al. (1991) But, are these programs effective?

(9) Finally, from the perspective of worldview theory, one must ask what would an examination of the other categories reveal since Nature represents only one of seven worldview categories (Cobern, 1991b)? For example, given the breadth of perspective students in this research brought to their understanding of Nature, do they also bring a diverse understanding of causality? Cobern (1989 & 1997) suggests they might. If so, do the students distinguish *scientific* causal explanations from (say) aesthetic or economic causal explanations within the domain of the natural world? Similarly, does science dominate the science teacher's causal under-

standing? Is causality interpreted differently by science teachers in different fields of science?

The assumption that underlies all of these questions, as well as the book, is the importance for science educators to understand the fundamental, culturally based beliefs about the world that students bring to class. Indeed, the underlying argument is that science education is successful only to the extent that science can find a niche in the cognitive and cultural milieu of students. Science education researchers are thus led to a principal question of cultural anthropology: What do people believe about the world and why? The "why" is crucial, for this indicates the question is about culture. Moreover, this question of culture is not only to be addressed to the lay public. It is equally important to ask these questions of the community of science and science education.

I further suggest that there is a very useful confluence of three disciplinary avenues emerging in science education. One is the avenue of cultural studies as represented by the present book, Allen (1995), Charron (1991), Cobern (1998, 1999), Costa (1995), George (1986, 1999), Jackson et. al. (1995), Larson (1995). Lowe (1997), McKinley (1997), Ogawa (1986, 1989, 1995, 1998), and Yakuba (1992, 1994), and others.[34] Cultural study implies the importance of learning in context and this is precisely the interest of research on situated cognition (e.g., Brown, Collins, and Duguid, 1989) and *contextualist learning* (Gordon, 1995), the second disciplinary avenue. The third avenue is connectionist or coherence epistemology (e.g., Bereiter, 1991). *Connectionism* serves to pull culture, cognition, and epistemology together for in the connectionist vision knowledge is a "vast network of interconnected elements" (Bereiter, 1991, p. 11) where the connections are at least as important as the elements if not more.

I thus return to Dewey's notion that all knowledge is continuous. Research along these avenues will seek to show how science education can be continuous with student culture while at the same time introducing students to the culture of modern science, to "guide students in the activity of 'scientificalizing' their world" (Lijnse, 1995, p. 192), without necessarily embracing the *extra*-scientific cultural accoutrements of various scientists, science teachers, and science writers.

[34] A "Workshop on Cultural Issues in Science Education" was held at the 1999 annual meeting of the National Association for Research in Science Teaching. The papers and discussion comments can be accessed at, http://www.ouhk.edu.hk/cridal/misc/iosteculture.html.

# *Chapter 10*
## Putting Things Together

As a science educator, I have wanted to learn something about the extent to which ninth graders will voluntarily enjoin scientific ideas (vis-à-vis other types of ideas) in a conversation only tacitly related to science. In this study, the findings were findings of diversity. Underneath the facade of demographic homogeneity, I found considerable conceptual diversity. Given the chance to talk freely about Nature, these students talked about *many* ideas. Science was only *one* of those many ideas and only sometimes mentioned. In contrast, interviews with science teachers show that the teachers almost immediately lapse into science talk. It is instructive to compare the diverse set of student ideas with an excerpt from a high school science teacher's narrative (as was done in Chapter 8). Ms Jackson was one of the more positivistic of the science teachers in this study and Figure 14 contrasts Ms Jackson with one of the students, Patricia. The contrasts are quite plain to see. However, diversity between students and teachers was not the only diversity observed in the study. The study also found remarkable diversity across teachers (as noted in Assertions 14 and 15, pages 85 and 88 respectively). Figure 15 contrasts comments from Mr. Hess, the other quite positivistic science teachers, with Mr. Bradford.

Mr. Hess and Ms Jackson offer a modernist view of science (Burtt, 1967) that strongly emphasizes reductionism. Their viewpoint is reminiscent of Francis Crick's Astonishing Hypothesis:

> You, your joys and your sorrows, your memories and your ambitions, your sense of personal identity and free will, are in fact no more than the behavior of a vast assembly of nerve cells and their associated molecules (Crick, 1994, p. 3).

Some have criticized this reductionist perspective as the "mythology of school science" (Smolic & Nunan, 1975) while for others it is "a candle in the dark" (Sagan, 1995). However one judges this modernist perspective on science or the reductionist nature of many traditional science curricula, it is clear that most of the ninth graders in the current study have a much different perspective. Unlike the science teachers, whether its Mr. Hess or Mr. Bradford, the students were not of their own accord focused on science. Except for Howard, the students conceptualized Nature as a composite of a number of different perspectives: aesthetic, religious, conservationist, and sometimes scientific.

| **Patricia** | **Ms Jackson** |
|---|---|
| God created the natural world. It has many characteristics: it's powerful, diverse, changeable, and beautiful (physically and emotionally). The Bible says God created the heavens and Earth, so I think that explains to me what Nature is.... Even though Nature is mysterious, everything is knowable but maybe not for us now, or even in the near future. The wonderment of the world increases knowledge through science, but is limited, due to its complexity.... | I think that Nature is predictable. I think that it is logical. I think that it is explainable. As scientists, we come up with laws of Nature or theories of Nature to be able to predict behaviors and therefore, based on what we know, and the experiments that we have done... there are certain patterns that we can find, and yet at times they can be very complex. But I think Nature, you can understand it... gravity is, and you throw a ball, then you can predict what is going to happen because it is logical.... If you were going to use lenses, and you know how light is going to go into the lenses, then let's say, with a telescope, you know that you need to use a lens to get the image and then the lens to invert the image, and so how you know how to use it, because of what it does. It is very logical to me. |

Figure 14

For two students, the conceptual variation was extreme. Art held remarkably pre-modern ideas for being a middle class, Anglo American student. He spoke of being much influenced by Native American thought and his account of Nature is reminiscent of an Australian Aboriginal view expressed by David Mowaljarlai.

> You have a feeling in your heart that you're going to feed your body this day, get more knowledge. You are looking at Nature and giving it your full attention. Seeing all its beauty. Your vision has opened and you start learning now.... When you touch them, all things talk to you, give you their story. It makes you really surprised.... You feel you want to get deeper, so you start moving around and stamp your feet – to come closer and recognize what you are seeing. (Mowaljarlai & Malnic, 1993, p. 53)

Another student spoke of Nature in explicitly anthropomorphic terms:

| Mr. Hess | Mr. Bradford |
|---|---|
| Nature is orderly and understandable. The tides and the rotation of the earth... That the planets and the stars are governed by physical forces and any deviations are simply because we have not yet discovered the other part of Nature's orderliness... As a science teacher I feel that with enough scientific knowledge all things are understandable... I think that the more we understand about matter itself, and the more we know about how to make things, the more predictable Nature will be. Scientific or reductionistic thinking is very powerful. I feel that once we know enough about the minutia of the world, breaking it down by using the scientific method, scientists tearing it apart and analyzing the parts of Nature and seeing how they interact, that we will be able to predict just about anything about Nature. | To me, the mysterious nature of nature is one of its better qualities. Things that are completely discovered are no longer interesting. For example, you have a cube of metal that everybody knows every single ingredient in it. Well, there is no mystery to it. There is nothing appealing about that anymore because there are no questions to ask about it. If everybody knows everything there is to know about that cube of metal, it looses appeal to me and I am sure that it looses its appeal to the people investigating it. If things don't have questions associated with them, there is no mystery. If there is nothing to ask about it anymore, it looses its intrigue, its interest. The mysteries of nature are hopefully unsolvable. I don't want to solve all the mysteries of nature. I hope nobody ever does. The appeal is like being lost out in the forest, so to speak. You want to be out there away from anything that is solved, you want to be in an environment where everything is still interesting to you. |

Figure 15

> **Betty:** Things in Nature have a consciousness. Since we are part of Nature and we have feelings, then Nature has feelings. Things in Nature have feelings. Plants, for example, scream when you pick a flower. That is something people don't realize or understand. The consciousness and the beauty of Nature are another type of powerful force.

Again, there is a striking parallel with traditional, non-Western thought. Consider the following comment of a Rock Cree individual of sub Arctic, Canada.

> Really [animals] resemble human beings, they differ only a little. They talk and eat and live like human beings. But you can only know this if you dream of them. It is excessively difficult to know these things. The animal doesn't want us to see it. Maybe we see a "small piece" of animals only. Only a "small piece" is what they show the people. (Brightman, 1993, p. 168)

Similar views concerning both plant and animal life can be found in Native American culture (e.g., Burbank, 1993), but such views are not what many would expect to find among the middle class Anglo students in this research. Moreover, such views certainly do not represent canonical science but represent a very different view of Nature and order within Nature. The contrast, thus, between Howard and Mr. Hess, the scientific materialists, and Art and Betty, the environmental mystics, could hardly be more noticeable. Of course, the example one has with Art and Betty, as noted above, is an extreme example. Other students differ from their science teachers and the science curriculum to a lesser degree. The point is that these differences are real and they have a real impact on the way students come to interpret, value and employ science. One result is that by current curriculum standards in science, there is little question which students are the more scientifically inclined and which are less. But what does it mean to be "scientifically inclined"?

## Border Crossing

Scientifically inclined? All major science curriculum or framework projects provide in one form or another an answer for this question.[35] Often those projects present a scientific worldview that emphasizes a rational, objective, material, technical and often utilitarian[36] perspective on Nature. Many science curriculum writers would

---

[35] Many science educators, including myself, consider Project 2061 to be the best of the American science curriculum guides. It is a worthy project that lays out the principles of good science teaching and the essential content of science appropriate at the school level. It does not, however, attend to the cultural issues raised in this book and in that sense Project 2061 is quite traditional.

[36] Although on this point, Project 2061 is better than most, it too concedes to the pragmatic:

> "Although the most powerful argument for improving the science education of all students may be its role in liberating the human intellect, much of the public discussion has centered on more concrete, utilitarian, and immediate justifications.... in our postindustrial society, there is a strong connection between how well a nation can perform and the existence of high-quality, widely distributed education. There is now a clear national consensus in the United States that all

likely argue that leaving out other perspectives should not be taken as a slight but a result of the proper demarcation of science vis-à-vis other domains of knowledge.[37] That could be true. It also could be true that like Mr. Hess, these are the terms that best describe the writers' primary vision of what the world is like, and they are oblivious to other terms. Either way, students are left to make their own accommodations with science. This was what Costa's (1995) study found about students enrolled in high school science. She characterized student attempts at accommodation as "border crossing" efforts and described several categories (also see Aikenhead, 1996; Aikenhead & Jegede, 1999; Cobern & Aikenhead, 1998). "Potential Scientists" cross borders into school science so smoothly and naturally that the borders appear invisible. Howard easily fits this description. "Other Smart Kids" manage their border crossing so well that few express any sense of science being a foreign subculture. This sounds like Patricia. Costa found that other students are unable to cross this border. These are the "Outsiders" who tend to be alienated from school itself and so border crossing into school science is virtually impossible. This description fits Art from this study, who eventually dropped out of school. One is thus led to the concern that the typical demarcation of science for school curricular purposes is far from benign, but inadvertently places hindrances in the path of students (such as Art and Betty) who probably could become "potential scientists" if given a proper chance.

Inadvertent hindrances to learning in the science classroom raise a question about what interactions can be better understood by explaining the interactions in terms of teacher/student conceptualizations of Nature, if indeed any? The suggestion here is that conflicts between student and teacher conceptualizations of Nature will have an adverse affect on the student's experience in the science classroom. For example, one might ask how well does an aesthetically minded person fare in Mr. Hess' science class?[38]

It is widely acknowledged that many students prefer biology to physical science. Women students in particular tend to prefer biology to physical science. Typically, one looks in the curriculum for the reasons for this difference. Is it possible that another significant factor could be teacher differences associated with subject area? An incident that occurred during one of the ninth grade interviews is suggestive of this very point. Ann is a good overall student and specifically a good science student. In her interview, Ann emphasized that Nature is something enduring and inclusive. Her sense of inclusiveness drew together knowledge of Nature, the natural beauty and purity of Nature, Nature as God's creation, and the

---

elementary and secondary school children need to become better educated in science, mathematics, and technology." (http://project2061.aaas.org/tools/sfaaol/Chap14.htm)

[37] See Smith and Scharman (1999) for a recent discussion of the demarcation of science from other domains of knowledge, and Cobern and Loving (2000a) for a discussion on defining science. Also see McNeill (1998) on the "scientific worldview."

[38] A pleasing side effect of this study was the impact of the student narratives on some of the science teachers. They decided to introduce time for open discussion in their science courses that would give students such as Ann the opportunity to ask questions about science, and talk about science, from a perspective of interest to the student.

conservation of Nature. In addition, she clearly spoke about Nature as something one can know about through science.

> **Ann**: Nature is knowable... We can learn to understand many things about Nature through personal experience, school and science. Science itself provides us with technology, which in turn increases our scientific know-ledge. Technology helps provide us with many wants, which, of course, increases our pleasure. It also uses resources.

This appreciation of science, however, is not where her narrative begins.

> **Ann**: Nature is something that is always out there and it will always be out there. Everything that exists is a part of Nature including you and me. To me, Nature is beautiful and pure because it is God's creation. Nature provides both aesthetic and emotional pleasure and I need it for self-renewal. I like to go where you can't see any influence by man. When I'm out in Nature, I feel calm and peaceful. It is a spiritual feeling and it helps me understand myself. I also get a spiritual feeling from Nature. Some-times, when I'm out in Nature and I have time to think, I start to wonder about things. This leads me to ask questions that I'd like to find answers to. The pleasure I get from Nature is enhanced by the mysteries I see in it.

Ann's conceptualization of the natural world has significant aesthetic and religious elements. Quite serendipitously during the interview, Ann mentioned her displeasure with the physical science class she was currently taking. I asked her to explain and she made it quite clear that the class was not about Nature as she had been discussing Nature. As one can see from the extended quote above, Nature for Ann is something friendly that you can joyously be part of. What impressed her about the physical science class was the teacher's warning about the dangerous chemicals that they would be handling during the course. The reasons for her displeasure with the class then became clear. She and her teacher had very different views of Nature.

One might be tempted to dismiss this young lady's aversion to dangerous chemicals as temporary and solely a result of insufficient conceptual understanding. She does not yet understand that there is danger in Nature, but with proper understanding and technique, this danger need not be viewed as a threat. That may be, but the question is how will this come about? Currently, Ann's aversion is rooted in an aesthetic sense of Nature that has more scope and force than her science teacher's assurances and explanations. It is critical that one notes that Ann's problem is not with science but with the context her science teacher chose to give science. Ann's mind not a map dominated by canonical scientific thinking. The thinking represents a *coherence* view of Nature where not one but several themes or large concepts have scope and force. A person accepts a belief or proposition as knowledge on the basis of coherence with other beliefs. Moreover, Aaron Wildavsky (1987) has argued persuasively from his cultural studies of people's political behavior that people are often able to make a broad range of quick assessments or decisions because these are consistent with a few strongly held cognitive elements. Unschuld (1995) comes to a similar conclusion in a study of how people make decisions about medical care.

Ann likewise is able to quickly assess her teacher's remarks about dangerous chemicals because these remarks are at odds, they do not cohere, with fundamental beliefs she holds about Nature. Ann has a sense of wonder about Nature that leads her to ask questions about Nature and thus adds to her understanding of Nature, including scientific and technological understanding. During the interview, Ann volunteered some information from science and technology as part of her discussion of what one can know about Nature. She showed an interest in scientific concepts but her foundation, the metaphysical frame that gives meaning to that interest, is in conflict with the classroom frame provided by the teacher. Ann has a sense of wonder about Nature but it is grounded in her fundamental view of Nature as beautiful and pure. If Ann continues in science, I suggest it will be because she has found her own way to accommodate what for her is an alienating view of science. On the other hand, she may well become one of Costa's (1995) "other smart people" who take and pass high school science courses only because this is required for college entrance a science class is simply one more hurdle one must jump in the school game. It is instructive to point out that Ann's science teacher at the time was Mr. Hess. Mr. Hess' conceptualization of Nature is dominated by a logical, reductionist view of knowledge and science. In contrast to Ann, his conceptualization of Nature is essentially monothematic.

As one can see, these are very different people but the difference could be characterized as that between an expert and a novice. Mr. Hess is the expert whose knowledge of science is "an environment in which there is located a collection of resources for knowing, understanding, and reasoning" (Lampert & Clark, 1990, p. 22). He knows when to draw on this environment and how to get around within it. The difference between the two, however, goes beyond the mere fact that one is an expert adult with considerable scientific education while the other is an adolescent and a novice with comparatively much less scientific education. Each has a very different orientation toward Nature, a different worldview, and one sees in their individual conceptualizations of Nature the roots of their actions in the classroom. The teacher's action is a rather matter-of-fact warning about dangerous chemicals. The student's action is a refusal to see this as legitimate talk about the natural world. Mr. Hess speaks quite naturally about the world using comfortable language for his lessons, all grounded in his fundamental view of reality. Similarly, Ann on entering the science classroom does not drop her *other* ideas, especially those with scope and force. Indeed, it is her background of other ideas that provides meaning for what she learns just as Mr. Hess' worldview provides meaning for what he teaches. This suggests that although Mr. Bradford, Mr. David, Mr. Hess, and Ms. Jackson are all considered effective science teachers, Ann would have preferred either Mr. Bradford or Mr. David to Mr. Hess or Ms. Jackson.[39] The implication of this research is that there are potentially influential factors in how teachers teach science that what would otherwise be considered no more than subtle background or

[39] In making this suggestion, I am aware that other factors must be assumed equal.

idiosyncratic differences among science teachers. As Helms (1998, p. 831) recently argued,

> teachers have more than a passing intellectual interest in their subject matter. In fact, dimensions of their identities are, to greater or lesser degrees, defined by it. Or, at the very least, they construct an identity in direct relation to science. Establishing the nature of science was not simply a philosophical exercise, but an attempt to understand more about themselves, to gain a richer understanding of who they are, why they do what they do, and, I argue, who they want to become.

That is to say, science is a personal matter with science teachers. It is not merely another subject. In contrast, one cannot expect most students to share this personal *scientific* bond. An episode from the research is suggestive of this point. This is an episode I describe earlier in Chapter 5 (p. 47-48). After all of the formal interviews had been conducted, I spoke with the students on another occasion. Ann and Howard were both present. I mentioned that during the interviews few of the students said much about science. Howard immediately protested that he had in fact said *very* much about science! Of course, he was correct. Ann protested differently. She said, "But you didn't ask anything about science!" Why would Howard (and all four science teachers) volunteer so much about science without being asked to do so, but Ann offers relatively little about science because she was not asked? Of course, one must ask, what is it about school science experience that science should have such a minimal impact on people's *everyday* thinking? At least in part, it must be due to something fundamentally personal that either does or does not place science in a cognitively central position in student thinking. In the perspective of this research, that fundamentally personal factor is worldview. The failure of the typical school curriculum is that it does not recognize science as a cultural phenomenon and hence the curriculum does not build cultural bridges – bridges that help students connect science with the other important aspects of their lives.

# Chapter 11

## Where is the Experience of Nature in School?[40]

An important observation in this study has been the disjunction between the students' experience of the world and the world as constructed in the science classroom. The disjunction is symptomatic of what Eger (1992, p. 342) called the "double distancing" between science and Nature that too often takes place in the science classroom. It is all the more unfortunate given the importance the majority of these ninth graders attached to personal experience with Nature and their strong aesthetic and environ-mental views that surely could be used to the advantage of science learning in the science classroom. As previously noted, environmentalism was the source of interest in science for several of the students in this study. Scientific literacy according to Hurd (1993) requires that students be able to independently use their knowledge of science in the everyday world. As Lijnse (1995) puts it, science education should thus seek to scientificalize a student's world. For the eventual scientific expert this means, "that as expertise is attained, a person restructures his/ her knowledge of the domain into a framework that is based on critical dimensions that facilitate the daily use of that knowledge" (Smith, 1992, p. 179). One has to wonder, however, if the future scientific expert and layperson both start at the same place? It could be that science education facilitates this cognitive restructuring process for all students; or, it could also be that science education is a natural selection device that selects for those students already amenable to a certain type of cognitive restructuring, as suggested by West (1996). For example, The *National Science Education Standards* opens with a quote from Richard Feynman that is intended to be inspiring.

> The world looks so different after learning science. For example, trees are made of air, primarily. When they are burned, they go back to air, and in the flaming heat is released the flaming heat of the sun which was bound in to convert the air into tree. [A]nd in the ash is the small remnant of the part which did not come from air, that came from the solid earth instead. These are beautiful things, and the content of

---

[40] This is a quote from McCarthy (1995, p. 10)

> science is wonderfully full of them. They are very inspiring, and they can be used to inspire others (p. viii).

Yes, the ideas of science can be inspiring but one wonders how many students would use the word "beautiful" where Feynman used it, and how many students would be inspired by the same ideas? After all, it was Richard Feynman (1995, p. 99) himself who remarked, "I had [science] in my blood from the beginning." Similarly, Howard from the present study would likely agree with Feynman, but it seems clear that Howard's fondness for science *preceded* his experiences with formal school science. Though based on anecdotal evidence, the suggestion here is that science education as now typically conceived is better at selecting for future scientific experts than it is at promoting public scientific literacy.

I noted earlier an interest in science classroom environmental change rather than a diagnostic-prescriptive approach to improving science teaching that essentially assumes that the environment is non-problematic. In the interest of promoting public scientific literacy, I suggest that classroom environment is a critical factor and that a critical classroom environmental problem is the radical isolation of school science from other disciplines and everyday thinking in general (Britzman, 1986). The assertions of this research suggest that science education does too little to help students integrate the important concepts of their own worlds with the important concepts of science. The assertions suggest that these students would benefit from a science classroom environment that put to use Dobzhansky's (1968, p. 242) insight that, "Knowledge gained from science is as necessary as it is by itself insufficient. It must be supplemented by the insights of poets, artists, mystics, and by religious experience." Indeed, these people are already in the classroom: Art the mystic; Ann the aesthetically minded; Patricia the Christian; Kevin the environmentalist; Jackie the Native American; and many more. As one of our teacher/researchers commented upon reading the student narratives, "what a wasted resource" – wasted because these are things that largely go untapped in the typical science classroom.

I have modestly suggested elsewhere how this might be done in the classroom (Cobern, 1991a & 1995; Cobern et al., 1995b; Cobern & Loving, 1998). Martin and Brouwer (1991, 1993) provide an excellent resource on science education for helping students develop a personal science. Flannery (1991) develops the idea of an aesthetic understanding of science and Poole (1995) handles the difficult issue of religious understandings of science. STS innovations (e.g., Cross & Price, 1992; Layton et al., 1993) and cultural studies (e.g., Rowe & Probst, 1995) are explicit attempts to change science curricula so that there is much greater interaction between science and the everyday world of students. What these authors have in common is an interest in promoting a science classroom environment that invites students to bring all of their important ideas to a dialogue with the important ideas of science.[41] This is not a suggestion for a bull session curriculum for the sharing of ignorance, but for what physicist David Bohm (1992, p. 16) called a dialogue,

---

[41] Also see Chambers (1999) and Levy (1999).

> The image this [dialogue] suggests is of a stream of meaning flowing among us... a flow of meaning in the whole group, out of which will emerge some new understanding... When everybody is sensitive to all the nuances going around, and not merely to what is happening in one's own mind, there forms a meaning which is shared.

The assertions of this research suggest to that just such a classroom dialogue is critical for the construction of scientific meaning within the everyday worlds of future (non-expert but literate) citizens.

Eleanor Duckworth has argued "that dealing in concepts-qua-nouns is barren – an example where someone can 'have' one of these things, and not be able to do much with it" (Duckworth, 1987, p. 51). The research reported in this book is part of an endeavor to better understand the concepts that have scope and force in student and teacher understandings of Nature. The research asked what are the science concepts that people use to conceptualize Nature and to what extent are they used? How are scientific ideas related to ideas from other disciplines? What is clear is that ninth grade students tend to discuss the natural world using several different perspectives, for example, aesthetic, conservationist, religious, and sometimes scientific. The group of ninth graders included some "science types." Other students likely learned their science as concepts-qua-nouns; but in my view, these and other students are not scientifically literate until the conceptual knowledge they have of science is meaningfully integrated into a cognitive framework that includes their everyday thinking. That, however, requires that educators come to know their students as persons. Duckworth (1987, p. 112) explained:

> Meaning is not given to us in our encounters, but it is given by us constructed by us, each in our own way, according to how our understanding is currently organized. As teachers, we need to respect the meaning our students are giving to the events that we share. In the interest of making connections between their under-standing and ours, we must adopt an insider's view: seek to understand their sense as well as help them understand ours.

Unfortunately, the insider's view involves perspectives too often ignored in the science classroom. Fourez (1988, p. 269) was once told by a colleague, "My course is scientific, period." Mitman, Mergendoller, Marchman, and Parker (1987, p. 626) reported from their research that, "the most striking results [were] that the sample teachers rarely or never went beyond science content in their instruction by trying to relate that content to other domains of scientific literacy and provide a larger context for the understanding of science facts and concepts." Perhaps then one should not be surprised by the reception given to a British proposal to place school science within a broader philosophical context. McCarthy (1995, p. 8) observed that science has always "explicitly and inevitably concerned itself with questions of existence, questions about the Nature and purpose of being, the Nature of matter." He was, however, surprised at his colleagues' response to the proposed curriculum revisions – revisions he thought unproblematic:

> Not our job... Nothing to do with us... These questions belong in [Religious Education], not science

they said (p. 7) and McCarthy asked himself incredulously,

> What does this indicate? Since when has science ceased to take an interest in questions of meaning, and purpose, beauty and mystery? Is there some connection between this blinkered world view and the widely attested loss of interest in science in schools? (p. 7)

I think so. I can easily imagine Art, Betty, and Simon wanting to know what science has to do with the meaning, purpose, beauty, and mystery of Nature. For these students science education and science must,

> undergo enrichment and de-formalization, getting cross-connected with the familiar phenomena of everyday life; and the familiar 'common sense' ideas not suppressed or declared wrong, but reconnected and re-constructed. (Hawkins, 1992, p. 219)

And, if science education is re-constructed so as to be de-formalized and cross-connected with the familiar phenomena of everyday life, then an answer will have been provided for the questions asked on page 4: Should science education seek to educate "scientific thinkers" in the pattern of the teachers? Or, should science education seek to foster sound science learning within the matrices of various cultural perspectives? Science education de-formalized and cross-connected will surely result in the second.

# *References*


Aidala, A. A. (1984). Worldviews, ideologies and social experimentation: Clarification and replication of "The Consciousness Reformation". *Journal for the Scientific Study of Religion, 23*(1), 44-59.

Aikenhead, G. S. (1996). Science education: Border crossing into the subculture of science. *Studies in Science Education, 27*, 1-52.

Aikenhead, G. S., & Jegede, O. J. (1999). Cross-cultural science education: A cognitive explanation of a cultural phenomenon. *Journal of Research in Science Teaching, 36*(3), 269-287.

Alcoze, T. (1991). Earth alive, circle whole. Earth Work: National Park Service, 1(7), 29-31, 33.

Allen, N. J. (1996). *'Voices from the bridge' Kickapoo Indian students and science education: A worldview comparison*. Unpublished doctoral dissertation, University of Texas, Austin, Austin, TX.

Allen, N. J., & Crawley, F. E. (1998). Voices from the bridge: Worldview conflicts of Kickapoo students of science. *Journal of Research in Science Teaching, 35*(2), 111-132.

Alters, B. J. (1997). Whose nature of science? *Journal of Research in Science Teaching, 34*(1), 39-55.

American Association for the Advancement of Science (AAAS). (1990). *Project 2061: Science for all Americans*. Washington, DC: American Association for the Advancement of Science, Inc.

American Association for the Advancement of Science (AAAS). (1993). *Benchmarks for science literacy*. Washington, DC: American Association for the Advancement of Science, Inc.

Arcury, T. A., Johnson, T. P., & Scollay, S. J. (1986). Ecological worldview and environmental knowledge: the "new environmental paradigm". *The Journal of Environmental Education, 17*(4), 35-40.

Atwater, M. M., Wiggins, J., & Gardner, C. M. (1995). A study of urban middle school students with high and low attitudes toward science. *Journal of Research in Science Teaching, 32*(6), 665-677.

Ausubel, D. P., Novak, J. D., & Hanesian, H. (1986). *Educational psychology: a cognitive view* (2d ed.). New York: Werbel & Peck.

Bagwell-Reese, M. K., & Brack, G. (1997). The therapeutic use of reframing and worldview in mental health counseling. *Journal of Mental Health Counseling, 19*(1), 78-86.

Baker, D., & Leary, R. (1995). Letting girls speak out about science. *Research in Science Teaching, 32*(1), 3-27.

Baker, D. A. (1998). *A study of the effect of culture on the learning of science in non-western countries.* Unpublished doctoral dissertation, Curtin University of Technology, Perth, Australia.

Baldwin, J. A. (1985). Psychological Aspects of European Cosmology in American Society: African and European Cultures. *Western Journal of Black Studies, 9*(4), 216-23.

Baldwin, J. A., & Bell, Y. R. (1985). The African self-consciousness scale: an africentric personality questionnaire. *Western Journal of Black Studies, 9*(2), 1-68.

Baldwin, J. A., & Hopkins, R. (1990). African-American and European-American cultural differences as assessed by the worldviews paradigm: An empirical analysis. *The Western Journal of Black Studies, 14*(1), 38-52.

Barr, J., & Birke, L. (1994). Women, science, and adult education: Toward a feminist curriculum? *Women's Studies International Forum, 17*(5), 473-483.

Belenky, M. F., Clinchy, B. M., Goldberger, N. R., & Tarule, J. M. (1986). *Women's ways of knowing: The development of self, voice, and mind.* New York: Basic Books, Inc.

Bereiter, C. (1991). Implications of connectionism for thinking about rules. *Educational Researcher, 20*(3), 10-16.

Bliss, J., & Ogborn, J. M. (1987). Knowledge elicitation. In J. R. Ennals (editor), *Artificial Intelligence: State of the Art Report* Vol. 15, Chap. 3, (pp. 207-262). New York: Pergamon Infotech.

Bohm, D. (1992). On dialogue. *Noetic Sciences Review,* (23), 16-18.

Brightman, R. (1993). *Grateful prey: Rock Cree human-animal relationships.* Berkeley: University of California Press.

Britzman, D. P. (1986). Cultural myths in the making of a teacher: Biography and social structure. *Harvard Educational Review, 56*(4), 442-456.

Brown, J. S., Collins, A., & Duguid, P. (1989). Situated cognition and the culture of learning. *Educational Researcher, 18*(1), 32-42.

Burbank, J. (1993). Yenaldlooshi: The shape-shifter beliefs of the Navajos. *The World & I, 8*(11), 263-271.

Burtt, E. A. (1967). *The metaphysical foundations of modern physical science.* London, UK: Routledge and K. Paul.

Carey, S. (1986). Cognitive science and science education. *American Psychologist, 41*(10), 1123-1130.

Carpenter, D. D. (1996). *Emergent nature spirituality: an examination of the major spiritual contours of the contemporary pagan worldview.* Albany, NY: State University of New York Press.

Catton, W. R. Jr., & Dunlap, R. E. (1978). Environmental sociology: A new paradigm. *The American Sociologist, 13,* 41-49.

Chambers, D. W. (1999). Seeing the world in a grain of sand: Science teaching in multicultural context. *Science & Education, 8*(6), 633-644.

Charron, E. H. (1991). Classroom and community influences on youths' perceptions of science in a rural county school system. *Journal of Research in Science Teaching, 28*(8), 671-687.

Cobern, W. W. (1989). Distinguishing science-related variations in the causal universal of college students' world views. Paper presented at the annual meeting of the *National Association for Research in Science Teaching* ERIC # 304346.

Cobern, W. W. (1991a). Introducing teachers to the philosophy of science. *Journal of Science Teacher Education, 2*(3), 45-47.

Cobern, W. W. (1991b). *World view theory and science education research, NARST Monograph No. 3.* Manhattan, KS: National Association for Research in Science Teaching.

Cobern, W. W. (1993a). College students' conceptualizations of nature: An interpretive world view analysis. *Journal of Research in Science Teaching, 30*(8), 935-951.

Cobern, W. W. (1993b). Contextual constructivism: The impact of culture on the learning and teaching of science. In K. G. Tobin (editor), *The practice of constructivism in science education* (pp. 51-69). Hillsdale, NJ: Lawrence Erlbaum Associates, Inc.

Cobern, W. W. (1995). Science education as an exercise in foreign affairs. *Science & Education, 4*(3), 287-302.

Cobern, W. W. (1996). Worldview theory and conceptual change in science education. *Science Education, 80*(5), 579-610.

Cobern, W. W. (1997). Distinguishing science-related variations in the causal universal of college students' worldviews. *Electronic Journal of Science Education, 1*(3), n.p.

Cobern, W. W. (Ed.) (1998). *Socio-cultural perspectives on science education: an international dialogue*. Dordrecht, The Netherlands: Kluwer Academic Publishers.

Cobern, W. W., & Aikenhead, G. S. (1998). Culture and the learning of science. In B. Fraser, & K. G. Tobin (editors), *International handbook of science education, Part Two* (pp. 39-52). Dordrecht, The Netherlands: Kluwer Academic Publishers.

Cobern, W. W., Gibson, A. T., & Underwood, S. A. (1995a). Everyday thoughts about nature: An interpretive study of 16 ninth graders' conceptualizations of nature - Working paper no. 127 of the Scientific Literacy and Cultural Studies Project (SLCSP). Paper presented at the annual meeting of the *National Association for Research in Science Teaching* ERIC #ED381401: San Francisco, CA.

Cobern, W. W., Gibson, A. T., & Underwood, S. A. (1995b). Valuing scientific literacy. *The Science Teacher, 62*(9), 28-31.

Cobern, W. W., Gibson, A. T., & Underwood, S. A. (1999). Everyday thoughts about nature: An interpretive study of 16 ninth graders' conceptualizations of nature . *Journal of Research in Science Teaching, 36*(5), 541-564.

Cobern, W. W., & Loving, C. C. (1998). The card activity: Introducing teachers to the philosophy of science. In W. McComas (editor), *The nature of science in science education: Rationales and strategies* (pp. 73-82). Dordrecht, Netherlands: Kluwer Academic Publishers.

Cobern, W. W., & Loving, C. C. (2000a). Defining 'science' in a multicultural world: Implications science education. *Science Education,* in press.

Cobern, W. W., & Loving, C. C. (2000b). *Enacted scientific worldviews: A case study of four high school science teachers: SLCSP Working Paper #139b.* Kalamazoo, MI: Scientific Literacy and Cultural Studies Project <http://www.wmich.edu/slcsp/SLCSP139/SLCSP139b.PDF>.

Costa, V. B. (1993). School science as a rite of passage. *Journal of Research in Science Teaching, 30*(7), 649-668.

Costa, V. B. (1995). When science is "another world": Relationships between worlds of family, friends, school, and science. *Science Education, 79*(3), 313-333.

Crick, F. (1994). *The astonishing hypothesis: The scientific search for the soul.* NY: Scribners.

Cronbach, L. J. (1975). Beyond the two disciplines of scientific psychology. *American Psychologist, 30*(2), 116-127.

Cross, R. T., & Price, R. F. (1992). *Teaching science for social responsibility*. Sydney, Australia: St. Louis Press.

Denzin, N. K., & Lincoln, Y. S. (editors). (1994). *Handbook of qualitative research*. Thousand Oaks, CA: SAGE Publications.

Dewey, J. (1976). *Experience and education*. New York: Collier Books/Macmillan Publishing Co., Inc.

DeWitt, C. B. (1998). *Caring for Creation: Responsible Stewardship of God's Handiwork*. Grand Rapids, MI: Baker.

Dijksterhuis, E. J. (1986). *The mechanization of the world picture: Pythagoras to Newton*. Princeton, NJ: Princeton University Press.

Dobzhansky, T. (1968). Teilhard De Chardin and the orientation of evolution: A critical essay. *Zygon, 3*(3), 242-258.

Duckworth, E. (1987). *"The having of wonderful ideas" and other essays on teaching and learning*. New York: Teachers College.

Dunlap, R. E., & Van Liere, K. D. (1978). The "new environmental paradigm": A proposed measuring instrument and preliminary results. *Journal of Environmental Education, 9*(4), 10-19.

Dzama, E. N. N., & Osborne, J. F. (1999). Poor performance in science among African students: An alternative explanation to the African worldview thesis. *Journal of Research in Science Teaching, 36*(3), 387-405.

Eger, M. (1992). Hermeneutics and science education: An introduction. *Science & Education, 1*(4), 337-348.

Emereole, H. U. (1998). World views portrayed by illiterate Botswana adults about common practices and phenomena. *Journal of the Southern African Association for Research in Mathematics and Science Education, 2*(1), 61-71.

Erickson, F. (1998). Qualitative research methods for science education. In B. Fraser, & K. G. Tobin (editors), *The international handbook on science education* (Vol. 2pp. 1155-1173). Dordrecht, The Netherlands: Kluwer Academic Publishers.

Fenstermacher, G. D. (1979). A philosophical consideration of recent research on teacher effectiveness. *Review of Research in Education, 6*, 157-185.

Fetterman, D. M. (1989). *Ethnography: Step by step*. Newbury Park, CA: Sage Publications, Inc.

Fetz, R. L., & Reich, K. H. (1989). World views and religious development. *Journal of Empirical Theology, 2*, 46-61.

Feynman, R. P. (1995). What is science? In D. K. Nachtigall (editor), *Internalizing physics: Making physics part of one's life - Eleven essays of Nobel laureates* (48 ed., Vol. 48pp. 99-112). Paris, France: United Nations Educational, Scientific and Cultural Organization.

Flake, C. L., Kuhs, T., Donnelly, A., & Ebert, C. (1995). Reinventing the role of teacher: Teacher as researcher. *Phi Delta Kappan, 76*(5), 405-407.

Flannery, M. C. (1991). Science and aesthetics: a partnership for science education. *Science Education, 75*(5), 577-593.

Foote, A. C. (1992). Rescued. *The World & I, 7*(11), 368-373.

Foster, J. W. (1991). Natural science and Irish culture. *Eire-Ireland; a Journal of Irish Studies, 26*(2), 92-103.

Foster, M. B. (1935). Christian theology and modern science of nature (I.). *Mind, XLIV*(176), 439-466.

Foster, M. B. (1936). Christian theology and modern science of nature (II.). *Mind, XLV*(177), 1-27.

Fourez, G. M. (1988). Ideologies and science teaching. *Bulletin of Science, Technology, and Society, 8*, 269-277.

Gallagher, J. J. (1991). *Interpretive Research in Science Education, NARST Monograph, Number 4*. Kansas State University: National Association for Research in Science Teaching.

Gardner, A. L., & Cochran, K. F. (1993). *Critical issues in reforming elementary teacher preparation in mathematics and science: Conference proceedings* Greeley, CO: Center for Research on Teaching and Learning.

Geertz, C. (1973). *The interpretation of culture*. New York, NY: Basic Books.

George, J. M. (1986). "Street science" in Trinidad and Tobago - Analysis and implications for teaching conventional science. In *Proceedings of the first regional consultation on science education research in Latin America and the Caribbean* . Trinidad.

George, J. M. (1999). World view analysis of knowledge in a rural village: Implications for science education. *Science Education, 83*(1), 55-75.

Gilligan, C. (1982). *In a different voice: Psychological theory and women's development*. Cambridge, MA: Harvard University Press.

Glacken, C. H. (1967). *Traces on the rhodian shore: Nature and culture in western thought from ancient times to the end of the eighteenth century*. Berkeley, CA: University of California Press.

Gordon, E. (1995). Culture and sciences of pedagogy. *Teachers College Record, 97*(1), 33-46.

Griffith, B. E., & Benson, G. D. (1994). Scientific thought as dogmatism. *International Journal of Science Education, 16*(6), 625-637.

Grumbling, V. O., Carter, J., Morgan, P., Lemons, J., & Saboski, E. (1991). Teaching nature literature in a science learning community: Merging "the two cultures". *CEA Critic, 54*(1), 47-51.

Gunstone, R. F. (1988). Learners in science education. In P. Fensham (editor), *Development and Dilemmas In Science Education* (pp. 73-95). New York, NY: The Falmer Press.

Harrison, P. (1999). Subduing the earth: Genesis 1, early modern science, and the exploitation of nature. *The Journal of Religion, 79*(1), 86-109.

Hawkins, D. (1992). Letter to the editor. *Science & Education, 1*(2), 219.

Hawkins, J., & Pea, R. D. (1987). Tools for bridging the cultures of everyday and scientific thinking. *Journal of Research in Science Teaching, 24*(4), 291-307.

Heath, S. B., & McLaughlin, M. W. (1994). Learning for anything everyday. *Journal of Curriculum Studies, 26*(5), 471-489.

Heller, P. M., & Finley, F. N. (1992). Variable uses of alternative conceptions: A case study in current electricity. *Journal of Research in Science Teaching, 29*(3), 259-275.

Helms, J. V. (1998). Science-and me: Subject matter and identity in secondary school science teachers. *Journal of Research in Science Teaching, 35*(7), 811-834.

Hiebert, P. G. (1976). *Cultural Anthropology*. Philadelphia, PA: J.B. Lippencott Company.

Holton, G. (1993). *Science and anti-science*. Cambridge, MA: Harvard University Press.

Horton, R. (1993). *Patterns of thought in Africa and the West: essays on magic, religion, and science* . Cambridge, UK: Cambridge University Press.

Hurd, P. D. (1993). Comment on science education research: A crisis of confidence. *Journal of Research in Science Teaching, 30*(8), 1009-1011.

Ibrahim, F. A. (1991). Contribution of cultural worldview to generic counseling and development. *Journal of Counseling and Development, 70*(1), 13-19.

Ibrahim, F. A. (1993). Existential worldview theory: transcultural counseling. J. McFadden (editor), *Transcultural Counseling: Bilateral and International Perspectives* . Alexandria, VA: American Counseling Association.

Ibrahim, F. A., & Kahn, H. (1987). Assessment of world views. *Psychological Reports, 60*, 163-176.

Jackson, A. P., & Sears, S. J. (1992). Implications of an Africentric worldview in reducing stress for African American women. *Journal of Counseling and Development, 71*(2), 184-90.

Jackson, D. F., Doster, E. C., Meadows, L., & Wood, T. (1995). Hearts and minds in the science classroom: The education of a confirmed evolutionist. *Journal of Research in Science Teaching, 32*(6), 585-611.

Jegede, O. J. (1997). School science and the development of scientific culture: A review of contemporary science education in Africa. *International Journal of Science Education, 19*, 1-20.

Jegede, O. (1998). The knowledge base for learning in science and technology education. P. Naidoo, & M. Savage (editors), *African science and technology education into the new millennium: Practice, policy and priorities* (pp. 151-176). Kenwyn, Republic of South Africa: Juta & Co Ltd.

Jenkins, E. W. (1992). School science education: towards a reconstruction. *Journal of Curriculum Studies, 24*(3), 229-246.

Jones, W. T. (1972). World views: Their nature and their function. *Current Anthropology, 13*(1), 79-109.

Kahle, J. B., & Meece, J. (1994). Research on gender issues in the classroom. In D. L. Gabel (editor), *Handbook of research on science teaching and learning* (pp. 542-557). New York: MacMillan Publishing Company.

Kawagley, A. O. (1995). *A Yupiaq worldview: A pathway to ecology and spirit*. Prospect Heights, IL: Waveland Press, Inc.

Kawagley, A. O., & Norris-Tull, D. N.-T. R. A. (1998). The indigenous worldview of Yupiaq culture: Its scientific nature and relevance to the practice and teaching of science. *Journal of Research in Science Teaching, 35*(2), 133-144.

Kawasaki, K. (1990). A hidden conflict between western and traditional concepts of nature in science education in Japan. *Bulletin of the School of Education, Okayama University*, (83), 203-214.

Kawasaki, K. (1996). The concepts of science in Japanese and Western education. *Science & Education, 5*(1).

Kearney, H. (1971). *Science and change, 1500-1700*. New York, NY: McGraw-Hill Book Company.

Kearney, M. (1984). *World view*. Novato, CA: Chandler & Sharp Publishers, Inc.

Keller, E. F. (1983). *A feeling for the organism: the life & work of Barbara McClintock*. San Francisco, CA: W. H. Freeman.

Kempton, W., Boster, J. S., & Hartley, J. A. (1995). *Environmental values in American culture*. Cambridge, MA: The MIT Press.

Kidwell, C. S. (1985). Native knowledge in the Americas. *OSIRIS* (Bruges, Belgium), 2nd series(1), 209-228.

Kilbourn, B. (1984). World views and science teaching. In H. Munby, G. Orpwood, & T. Russel (editors), *Seeing curriculum in a new light*. Lanham, MD: University Press of America.

Knight, L. U. (1996). *The voluntary human extinction movement (VHEMT).*

Knopf, R. C. (1987). Human behavior, cognition, and affect in the natural environment. In D. Stokols, & I. Altman (editors), *Handbook of environmental psychology* (pp. 783-825). New York, NY: John Wiley & Sons, Inc.

Kraft, C. (1974). Ideological factors in intercultural communication. *Missiology, 2.*

Kraft, M. G. (1978). *Worldview and the communication of the Gospel*. Pasadena, CA: William Carey Library.

Kvale, S. (1983). The qualitative research interview: A phenomenological and a hermeneutical mode of understanding. *Journal of Phenomenological Psychology, 37*, 171-196.

Lampert, M., & Clark, C. M. (1990). Expert knowledge and expert thinking: A response to Floden and Klinzing. *Educational Researcher, 19*(5), 21-23, 42.

Larson, J. O. (1995). Fatima's rules and other elements of an unintended chemistry curriculum. Paper presented at the annual meeting of the *American Educational Research Association* San Francisco, CA.

Lassiter, I. H. I. (1993). *Ways of knowing among college nonscience majors: a world view investigation*. Unpublished doctoral dissertation, Georgia State University, Atlanta, GA.

Lawrenz, F., & Gray, B. (1995). Investigation of worldview theory in a South African context. *Journal of Research in Science Teaching, 32*(6), 555-568.

Layton, D. (n.d.). Inarticulate science? Or, does science understand its publics as well as its publics understand science? *MSTEnews, 3*(2), no page numbers.

Layton, D., Jenkins, E., Macgill, S., & Davey, A. (1993). *Inarticulate science? Perspectives on the public understanding of science and some implications for science education*. Nafferton, Driffield, East Yorkshire: Studies in Education Ltd.

Lederman, N. G. (1992). Students' and teachers' conceptions of the nature of science: A review of the research. *Journal of Research in Science Teaching, 29*(4), 331-359.

Lederman, N. G. (1995). Suchting on the nature of scientific thought: Are we anchoring curricula in quicksand? *Science & Education, 4*(4), 371-377.

Lee, O. (1999). Science knowledge, world views, and information sources in social and cultural contexts: Making sense after a natural disaster. *American Educational Research Journal, 36*(2), 187-219.

Levitt, N., & Gross, P. R. (1994). The perils of democratizing science. *The Chronicle of Higher Education, XLI*(6), B1-B2.

Levy, S. (1999). To see the world grain of sand. *Educational Leadership, 57*(3), 70-75.

Lewis, B. F. (1996). *The influence of world view on African American college students' decisions to study science: An interpretive investigation of three cases*. Unpublished doctoral dissertation, The Florida State University, Tallahassee, FL.

Lewis, B. F. (1998). A review of worldview research in education: Analyzing theory, methods and applications. Paper presented at the annual meeting of the *American Educational Research Association*. San Diego, CA.

Lewis, C. S. (1960). *Studies in words, Second Edition (reprinted in 1994)*. Cambridge, UK: Cambridge University Press.

Lijnse, P. L. (1995). Developmental research as a way to an empirically based didactical structure of science. *Science Education, 79*(2), 189-199.

Lincoln, Y. S., & Guba, E. G. (1990). *Naturalistic inquiry*. Newbury Park, CA: Sage.

Loving, C. C., & Foster, A. (1998). The religion-in-the-science-classroom issue: Seeking graduate student conceptual change. Paper presented at the annual meeting of the American *Educational Research Association* San Diego, CA.

Lowe, J. A. (1995). The impact of school science on the world-view of Solomon Island students. *Prospects, XXV*(4), 653-667.

Lowe, J. A. (1997). Scientific concept development in Solomon Island students: a comparative analysis. *International Journal of Science Education, 19*(7), 743-759.

Lynch, P. P., & Jones, B. L. (1995). Student's alternative frameworks: Towards a linguistic and cultural interpretation. *International Journal of Science Education, 17*(1), 107-118.

Lythcott, J. (1991). The nature of essential knowledge bases for science teachers. *Teaching Education, 3*(2), 41-55.

Martin, B. E., & Brouwer, W. (1991). The sharing of personal science and the narrative element in science education. *Science Education, 75*(6), 707-722.

Martin, B. E., & Brouwer, W. (1993). Exploring personal science. *Science Education, 77*(4), 441-459.

Marton, F. (1988). Phenomenography: Exploring different conceptions of reality. In D. M. Fetterman (editor), *Qualitative approaches to evaluation in education: The silent scientific revolution* (pp. 176-205). New York, NY: Praeger.

Mattson, S. A. (1997). *When world views collide: Conflicting beliefs about assessment in an interdepartmental effort to develop a biology course for prospective elementary teachers*. Unpublished doctoral dissertation, The Florida State University.

McCarthy, K. (1995). Science: Power or wisdom? *School Science Review, 76*(276), 7-22.

McKinley, E. (1997). Science education from the margins: the uneasy selfhood of a postcolonial woman. Paper presented at the annual meeting of *ASERA* Adelaide, Australia: ASERA.

McNeill, W. H. (1998). History and the scientific worldview. *History and Theory, 37*(1), 1-13.

Merchant, C. (1989). *The death of nature: Women, ecology, and the scientific revolution*. San Francisco, CA: Harper & Row.

Miller, J. D. (1987). The scientifically illiterate. *American Demographics, 9*(9), 26-31.

Mitman, A. L., Mergendoller, J. R., Packer, M. J., & Marchman, V. A. (1987). Instruction addressing the components of scientific literacy and its relation to student outcomes. *American Educational Research Journal, 24*(4), 611-633.

Mowaljarlai, D., & Malnic, J. (1993). *Yorro yorro: Aboriginal creation and the renewal of nature - Rock paintings and stories from the Australian Kimberley*. Rochester, VT: Inner Traditions.

NAEP. (1979). *Three assessments of science, 1969-1977: Technical summary*. Washington, DC: Education Commission of the States.

Nakamura, H. (1980). The idea of nature, east and west. In M. J. Adler (editor), *The great ideas today, 1980* (pp. 234-304). Chicago: Encyclopaedia Britannica, Inc.

National Research Council. (1996a). *Mathematics and science education around the world: What can we learn?* Washington, DC: National Academy Press.

National Research Council. (1996b). *National science education standards*. Washington, DC: National Academy Press.

Nevers, P., Gehard, U., & Billman-Mahecha, E. (1997). Patterns of reasoning exhibited in response to moral dilemmas involving plants, animals and ecosystems. *Journal of Moral Reasoning, 26*(2), 169-186.

Noddings, N. (1993). *Educating for intelligent belief or unbelief.* New York: Teachers College Press.

Ødegaard, M. (in progress). *Tentative Title: Science Education and the Feminist Worldview.* Unpublished doctoral dissertation, University of Oslo, Norway, Oslo, Norway.

Ogawa, M. (1986a). A preliminary study on a key concept "nature" involved in science education. *The Bulletin of the Faculty of Education, Ibaraki University* (Educational Sciences) (In Japanese With English Abstract), (35), 1-8.

Ogawa, M. (1986b). Toward a new rational of science education in a nonwestern society. *European Journal of Science Education, 8,* 113-119.

Ogawa, M. (1989). An introductory study on traditional views of nature: In the context of pursuing rationale for science education. *The Bulletin of the Faculty of Education, Ibaraki University (Educational Sciences) (In Japanese With English Abstract),* (38), 47-58.

Ogawa, M. (1995). Science education in a multiscience perspective. *Science Education, 79*(5), 583-593.

Ogawa, M. (1998). A cultural history of science education in Japan: an epic description. In W. W. Cobern (editor), *Socio-cultural perspectives on science education: an international dialogue* (pp. 139-161). Dordrecht, The Netherlands: Kluwer Academic Publishers.

Ogunniyi, M. B., Jegede, O. J., Ogawa, M., Yandila, C. D., & Oladele, F. K. (1995). Nature of worldview presuppositions among science teachers in Botswana, Indonesia, Japan, Nigeria, and the Philippines. *Journal of Research in Science Teaching, 32*(8), 817-831.

Otijele, P. Y. (1991). Understanding the African worldview: a religious perspective. *Ogbomoso Journal of Theology,* 1-16.

Pauka, S. *The influence of culture on the learning of scientific concepts by students in Papua New Guinea.* Unpublished doctoral dissertation, Curtin University of Technology, Perth, Australia.

Perreault, G. (1979). Futuristic world views: Modern physics and feminism: Implications for Teaching/Learning in Higher Education. Paper presented at the *Annual Conference of the World Future Society* ERIC # 184 016.

Piaget, J. (1963). *The child's conception of the world.* Paterson, New Jersey: Littlefield, Adams & Co.

Piaget, J. (1971). *Biology and knowledge: An essay on the relations between organic regulations and cognitive processes,* translated by B. Walsh. Chicago: University of Chicago Press.

Polkinghorne, J. (1996). *Beyond science*. Cambridge, UK: Cambridge University Press.

Poole, M. W. (1995). *Beliefs and values in science education*. Philadelphia: Open University Press.

Postman, N. (1985). *Amusing ourselves to death: Public discourse in the age of show business*. 375 Hudson St., NY, NY 10014: Penguin Books USA Inc.

Proper, H., Wideen, M. F., & Ivany, G. (1988). World view projected by science teachers: a study of classroom dialogue. *Science Education, 72*(5), 542-560.

Radhakrishnan, S. (1967). *Religion in a changing world*. London, UK: George Allen & Unwin, Ltd.

Ramorogo, G. J. (1998). *Effects of exemplary teaching and learning materials on students' performance in biology*. Unpublished doctoral dissertation, University of the Western Cape, South Africa.

Redfield, R. (1952). The primitive world view. *American Philosophical Society, Proceedings, 96*, 30-36.

Ross, R. (1962). *Symbols and civilization: science, morals, religion, art*. New York: Harcourt, Brace & World, Inc.

Roth, W. M., & Alexander, T. (1997). The interaction of students' scientific and religious discourses: Two case studies. *International Journal of Science Education, 16*(2), 125-146.

Rowe, R., & Probst, C. (1995). Connecting with local culture. *Educational Leadership, 53*(1), 62-64.

Rudner, L. M., Song, T., Treacy, M. E., & Pike, L. (1984). *The national assessment of educational progress information retrieval system (NAEPIRS), Version 1.15*. Washington, DC: National Institute of Education, computer system.

Rutherford, F. J. (1987). The character of elementary science. *Science and Children, 24*(4), 8-11.

Sagan, C. (1995). *The demon-haunted world: science as a candle in the dark*. New York: Random House.

Shumba, O. (1999). Relationship between secondary science teachers' orientation to traditional culture and beliefs concerning science instructional ideology. *Journal of Research in Science Teaching, 36*(3), 333-355.

Shymansky, J. A., & Kyle, W. C. Jr. (1992). Establishing a research agenda: The critical issues of science curriculum reform. *Journal of Research in Science Teaching, 29*(8), 749-778.

Simon, Y. (1970). *The great dialogue of nature and space*. Albany, NY: Magi Books, Inc.

Slay, J. (1999). The Nature of Nature: Chinese Culture and Science Education. Paper presented at the annual meeting of the *National Association for Research in Science Teaching* .

Smith, M. U. (1992). Expertise and the organization of knowledge: Unexpected differences among genetic counselors, faculty, and students on problem categorization tasks. *Journal of Research in Science Teaching, 29*(2), 179-205.

Smith, M. U., & Scharmann, L. C. (1999). Defining versus describing the nature of science: A pragmatic analysis for classroom teachers and science educators. *Science Education, 83*(4), 493-509.

Smolicz, J. J., & Nunan, E. E. (1975). The philosophical and sociological foundations of science education: the demythologizing of school science. *Studies in Science Education, 2*, 101-143.

Snively, G. (1990). Traditional native Indian beliefs, cultural values, and science education. *Canadian Journal of Native Education, 17*(1), 44-59.

Snively, G., & Corsiglia, J. (2000). Discovering indigenous science: Implications for science education. *Science Education*, in press.

Spradley, J. (1979). *The ethnographic interview*. New York, NY: Holt, Rinehart and Winston, Inc.

Stake, R. E. (1988). Case study methods in educational research: Seeking sweet water. In R. M. Jaeges (editor), *Complementary methods for research in education* (pp. 253-269). Washington, DC: AERA.

Stillman, C. W. (1977). On the meanings of "nature". Paper presented at the symposium on *Children, Nature, and the Urban Environment: Proceedings of a Symposium-Fair. USDA Forest Service General Technical Report NE-30* (pp. 25-30). Upper Darby, PA: Forest Service, US Department of Agriculture Northeastern Forest Experiment Station.

Strauss, A. L. (1987). *Qualitative analysis for social scientists*. New York, NY: Cambridge University Press.

Tervino, J. G. (1996). Worldview and change in cross-cultural counseling. *Counseling Psychologist, 24*(2), 198-215.

The New York Times. (1995 April). NY kids awed by outdoors: `Sweat' on horses; `woods never end'. *The Arizona Republic*, p. A31.

Theocharis, T., & Psimopoulos, M. (1987). Where science has gone wrong. *Nature, 329*(6140), 595-598.

Thomas, K. (1983). *Man and the natural world: A history of the modern sensibility*. New York: Pantheon Books.

Tobin, K. G. (editor). (1993). The practice of constructivism in science education. Hillsdale, NH: Lawrence Erlbaum Associates.

Unschuld, P. U. (1995). Plausibility or truth? An essay on medicine and world view. *Science in Context, 8*(1), 9-30.

Waldrip, B. G., & Taylor, P. C. (1999). Permeability of students' worldviews to their school views in a non-western developing country. *Journal of Research in Science Teaching, 36*(3), 289-303.

Wallace, A. F. C. (1970). *Culture and personality*. New York: Random House.

Watanabe, M. (1974). The conception of nature in Japanese culture. *Science, 183*(4122), 279-282.

West, P. (1996). *Basic college science courses filter out most students*. Washington, DC: National Science Foundation.

Whatley, M. H. (1989). A feeling for science: Female students and biology texts. *Women's Studies International Forum, 12*(3), 355-362.

White Jr., L. (1967). The historical roots of our ecological crisis. *Science, 155*(3767), 1203-1207.

Wildavsky, A. (1987). Choosing preferences by constructing institutions: A cultural theory of preference formation. *American Political Science Review, 81*(1), 1-21.

Wilson, B. (1981). The cultural contexts of science and mathematics education: Preparation of a bibliographic guide. *Studies in Science Education, 8*, 27-44.

Witherspoon, G. (1974). The central concepts of Navajo world view (I). *Linguistics*, (119), 41-59.

Witherspoon, G. (1975). The central concepts of Navajo world view (II). *Linguistics*, (161), 69-88.

Wong, E. D. (1995). Challenges confronting the researcher/teacher: Conflicts of purpose & conduct. *Educational Researcher, 24*(3), 22-28.

Woolnough, B. E. (1989). Faith in science. *School Science Review, 70*, 133-137.

Yakuba, J. M. (1992). Indigenising the science curriculum in Ghana through the science in Ghana society project. *Science Education International, 3*(3), 14-19.

Yakuba, J. M. (1994). Integration of indigenous thought and practice with science and technology: A case study of Ghana. *International Journal of Science Education, 16*(3), 343-360.

Young, R. V. (1974). Christianity and ecology. *National Review, XXVI*(51), 1454-1458, 1477, 1479.

# *Appendix A*

## Student Narratives on Nature

### Alice

How could someone say 'Nature is just there' is beyond me because its everywhere. Nature is everywhere. It can't be something that's just there. We can't go a day without using something from nature. We depend on Nature for everything: material items, resources, ideas and pleasures. If we as humans aren't careful we are going to ruin the one thing that we need to survive. I see many sides to nature. It is material...full of resources: It is living...it can be hurt. I think Nature is very dangerous. I think Nature is very beautiful. It can be beautiful and peaceful but also dangerous and frightening. A tornado, for example, can be beautiful and mysterious in its power and at the same time ugly in the damage it can do. Nature has a predictable, understandable side to it, but also, an unpredictable, uncontrollable side. I want to be a scientist. Nature is very important to the world of science. Through science we understand many of the patterns in nature; food webs, weather patterns, how the solar system works, etc. We need to know more about Nature and we keep studying it to find out how things work and to discover ways that different things affect each other. However, while science can work to increase our knowledge and understanding of nature, it is all people in the world that must act responsibly to help solve the problems we've created in nature. It will be hard for scientists to study Nature in the future if we just keep destroying it. We use a lot of material that Nature gives. Nature is full of resources ... rocks, chemicals.... gas, oil... trees, animals....and air and water. Exploitation is a problem. It is frustrating that many people, except scientists, don't care. Not all of the problems are correctable. Although Nature is somewhat restorable, once plants and animals are extinct that's final. Is Nature doomed? Yes, possibly! For lack of caring! I'm not sure about the connection between God and nature. While I am a Christian, I also believe that science has proved wrong, many of the things in the Bible. Yet, I do think that there is a purpose for our existence and God is behind it. Science can explain how things work but there are many why questions that science doesn't answer. I have a very strong response to nature. I'm one of those people that is always thinking about Nature in everything I do or want to do. While I see beauty in Nature and that is very important to me I think it is far more important that we put out thoughts towards protecting Nature at this time. My own personal enjoyment is secondary to that need. Because Nature is so important to us, it is sacred. We have to have it to live. That's kind of a big deal.

**About Alice:**

She is an Anglo-American girl that likes to be involved in everything. She is friendly, open, outgoing, and a competent organizer. Consequently, she is welcome in whatever activity she shows an interest. Teachers like and appreciate her a lot and she gets a great deal of positive support from them. She has a very individualistic personality and although she is quite well liked, she doesn't have many close friends. She dresses with her own unusual flair and seems to deliberately keep herself away from the mainstream of students. At times Alice has discussed a less than happy home situation but has been very clear that she doesn't chose to feel sorry for herself because of it. She would rather build a good life for herself based on a strong education and move out of her home as soon as she is able. Alice's father and stepmother have a weed-control business and Alice expresses some discomfort about the damage that they may be causing to the environment. She also speaks of the camping trips and fishing that she likes to do with her family but can't remember the last time that they went. Academically, Alice is a conscientious student that works hard for her "A" grades. She has had Introduction to Physical Science, Biology and Chemistry. She is very active in the school drama program as well as DECA, Key Club, and has helped with several art and science projects. Out of school, Alice has participated in the Renaissance Festival and she likes to sing and dance. Her goals are to attend 4 years of college and then to get a master's degree in either acting or biology related field. Alice has a 3.00 GPA and a class rank of 51/144.

**Simon**

Although I've thought a little bit about the natural world, I don't really understand a lot of things. I suspect that much of Nature isn't meant to be understood. Because Nature lacks order and is often unpredictable, it is often unexplainable. Some things like weather and ocean patterns can be predicted but many dangerous things might not be predicted - earthquakes and natural disasters for example. Animals also do things that we don't understand and can't explain. Some aspects of Nature are knowable and it is important that we learn more about it. What we learn comes from both school and personal experience. Our lack of understanding of Nature has caused us to exploit our natural resources. Ultimately we are causing permanent damage because of such things as overpopulation, oil spills, cutting down trees, pollution, etc. Possibly we are doomed. We might be able to do some restoration that might help solve some of our problems. I really enjoy being out in nature. It gives me good feelings. I like walking around, climbing mountains, watching a deer drink out of a river and things like that. I think about Nature and you could say I'm in touch with nature. Though I understand only a little about it, I like the mystery of not understanding everything. It adds to the beauty. Nature can be peaceful, with calm breezes, lots of nice trees and no trash. I also have some religious feelings about nature. Not necessarily those of any one particular religious group. I do think that some God created the earth. This confuses me also. I'm not entirely sure of my beliefs but I do think that a god created the earth.

**About Simon**

Simon is an Anglo-American boy who is a puzzle to his teachers. Although he tests quite well on achievement tests, he is very unsuccessful in school. His GPA is only 1.17 with a class rank of 129/144. He has great difficulty organizing his thoughts both orally and in writing. This was very clear in his interviews. He became very fatigued during the interviews and was visibly distressed when asked to repeat or develop any of the thoughts that he alluded to. On the other hand, Simon is interested in the subjects discussed in his science classes, he will ask questions after class although rarely in class and has improved during the past two years in turning in papers, etc. Simon frequently has a disheveled appearance, as does his backpack, his books and papers. One of his teachers had occasion to go to his home where he lives with his mother and sister. It was in such a state of disorder that the teacher chose not to go inside. Simon is a very personable, somewhat shy young man. He has good friends that he spends time with at school. As a group they seem somewhat immature but certainly aren't unusual for their age. Simon is well liked and respected by his friends. He is kind and courteous, doesn't use foul language, etc. Simon's mother is a teacher. He intends to pursue a profession in the military, become a police officer or "work as a manager for my brother". He has a vague plan to go back East and go to college but nothing that has been clearly defined. He has taken three science classes in high school with limited success. He clearly states that he does not like to study. He is not involved in any extra-curricular activities at school and when asked what he likes to do away from school could only offer that he participates in those activities that he thinks are fun, but he couldn't think of any specifics at the time of our conversation. He did say that he likes to watch football and would like to fish. Simon also could think of no activities that he does with his family.

**Howard**

Nature is everything around us, both living and non-living, that we interact with. This includes outer space. Nature is also the ways and means of how things operate. The meaning of our lives is to interact with nature. That is our purpose. I think that Nature can be fully known because it is logical. We don't know or understand all of it yet but as time goes on we will understand more and more. Most things about Nature are somewhat orderly or have a pattern to them. Because of this, the study of science allows us to explain what is going on in nature. The orderliness lets us predict many things that are going to happen, like the weather, for example. Sometimes Nature seems chaotic but that is mostly because our knowledge is incomplete and therefore our understanding is limited. Because people are naturally curious, they pursue studies of Nature whether they are actively involved in Science or simply following up on a personal interest. I think that everything can be explained by science. Matter, both living and non-living, and what it does follows basic laws. Things like the Law of Conservation of Mass, reproductive cycles of plants and animals, convection currents and ecosystems can be understood if the laws of science are studied. The study of science can lead to some hard questions, for example, Why orbits go the direction that they do? In the future things like this will be fully known. Nature provides us with many resources. Energy, shelter, food and water all come from nature. Scientific studies will allow us to use more of Nature to our advantage. Our exploitation has caused pollution. Humans have definitely influenced the natural world by building cities and communities but I think that Nature is restorable. It can be restored by natural processes if left alone. We are also able to restore it somewhat by conservation efforts. We need to be careful, though, of environmental extremism. Nature also has a beautiful side to it. It can be very peaceful. I find Nature to be peaceful when I'm hiking up a mountain or something like that. But I also find it peaceful when I'm just walking around at night sometimes. I am a religious person but I also try to take things as they appear to me so I don't believe that everything in Nature has a spiritual side to it. I do enjoy being out in Nature and thinking about different aspects of it. There is a beauty in all the ideas that there are in Nature to learn about.

**About Howard:**

Howard is an Anglo-American. One of his teachers called him a "science information gatherer." He likes to read about science, talk about science with his friends at lunch, discuss science with his teachers, watch science movies and do any sort of science related activities as long as they don't involve homework or school assignments. He maintains a "B" average in his science classes with a great deal of encouragement from his teachers to turn in his papers. He does not do as well in his other classes because, in his words, "they don't matter as much". Howard is clearly an intelligent young man, able to think his own thoughts and not embarrassed to discuss them and measure them against those of someone more knowledgeable than he. He also weighs what is said to him and does not automatically assume that what is taught is correct because it comes from an "authority". Howard is also a pleasure to have as a student. He is courteous, thoughtful and full of ideas. Howard's mother is a statistician and his father works for a computer company. Although his father has not come to the high school that Howard attends, his mother is a frequent visitor. She is one of Howard's best supporters. She takes visible pride in his intelligence and encourages him to explore new ideas. Howard speaks positively of his family. They go to church together and he has been quite active in his church. He is in the choir, has been a cantor and an alter server. He also participates in the church youth group. Howard intends to go into computer programming but enjoys the high school chorus and would like to keep music in his life. Howard is enrolled in his third science class in high school and plans to take two more his senior year.

After school, Howard participates in the drama club both acting and working on the lights, the PEP band and the school chorus. Away from school, his special interests are composing music, writing short stories and poetry, computer programming and drawing. He plans to attend a state university and probably graduate school. Howard has a GPA of 2.57 and class rank of 76/144.

## Art

No matter what we humans do we're still natural and we're part of the natural world. I believe that man does not stand separate from Nature but is part of it, including space, planets, oceans, living organisms and non-living things. Without the things we get from Nature, we could not enjoy the lifestyle that we have today. Nature is vulnerable to our influence and as our population has grown, Nature has been affected. Man has changed the natural world by exploiting its resources and polluting the environment. We have depleted the rain forests and changed the balance of natural things. I am concerned about the pollution we have caused, the things we have lost from the rain forest without knowing they are there, the damage we have done to Earth's water. Now we don't have the opportunity to gain knowledge from what we already have destroyed. Nature is a source of knowledge. That's a resource. Our ability to increase our knowledge and understanding of Nature will allow us to correct some of the damage that we have done. Nature is not doomed. Although we will probably not be able to restore things to their original state man has the capability of making positive changes to the natural world as well as negative ones. Nature is not doomed. It can always take a different course and we are part of that course change. At the present time our knowledge of the natural world is limited. Many things that we perceive to be complex and confusing because we don't understand them are actually quite simple and orderly. The construction of a spider web, for example, is quite a complicated operation to us but to the spider building the web it is a simple procedure. As we gain in understanding of the diversity and power of nature, we will understand the perfect balance of everything in nature. We will also begin to understand our place within nature. It is more important to have a spiritual understanding of Nature than just scientific knowledge. That understanding can't be gained from school. You have to spend time in Nature and learn to feel it. Than you will understand it. There is a spiritual aspect to Nature to many people. I find it quite beautiful, especially when I am looking out at a part of the natural world that is untouched by man. I appreciate the diversity of plants and animals in nature. Animals are very important to me, I can feel things through animals. I enjoy watching them and learning about them. I understand why Nature is sacred to some groups of people. It is sacred to me. The American Indian culture has the kind of understanding for Nature that encourages preservation rather than destruction. Scientists, also, are people that understand the need to preserve and protect. Some scientists and Indian cultures understand their role on Earth. They do not think of themselves as superior beings and don't feel that they have a right to go around destroying nature. They leave it the way it is because Nature was made a certain way and it is supposed to be kept that way.

Unfortunately scientists and scientific knowledge are also increasing our tendency to pollute, destroy and clutter up the earth and space. They are trying to destroy it and study it at the same time. I believe Nature needs to be protected. Everything has a purpose that is put on this planet. We have a purpose here also. Maybe it is to preserve nature, have fun and enjoy the spiritual quality instead of destroying it. It is our responsibility to know and understand human impact on the fragile, easily tampered with balance of the natural world. Nature is something felt! Understanding would be the most valuable key to the future of the world.

**About Art:**

Art is an Anglo-American who has been in and out of public high school and a private "individualized" high school with no success in either. Art has a class rank of 144/144 with no GPA (zero) because he has accumulated few credits. He claims that he has completed several preliminary classes in community college and will be leaving to enroll there quite soon. Meanwhile he is completely disengaged from learning, professing that reaching an "understanding" with the world is far more important. An example is his refusal to learn about a microscope in biology class because man would be so much better off without technology. Art also has considerable art talent but is unable to focus himself enough to accomplish much with it. He admits to using drugs but "only pot, and only on weekends" – but he says he has "fried his brain" and that it will take him a while to get over it. Art is a charmer. He speaks with concern over environmental issues, man's inhumanity to man, his own disinterest in dealing with a world that has few values, etc. He frequently voluntarily offers to complete tasks for people that he feels are sympathetic to him but there is rarely follow-through. Both of Art's parents are insurance agents. He prefers to be a reverend or a peacemaker. When asked about activities that his family does together his answer was "we take part in the separation and aren't concerned for each other's needs. Our personal material needs blind our feelings for each other." His immediate goal is to drop out of high school and attend community college. His special interests are mountain climbing, camping, anything mind-expanding, anything that can be "understood rather than known". He says,

> "I am lost. Hopelessness creeps around every corner. The past can't be accepted. Lies follow my conscience when at school. Participation in this bureaucracy disgusts me in every way."

**Samantha**

Even though Nature is complex and hard to understand I think Nature is inspirational. It has a powerful effect on me. Words like beautiful, powerful, pure and peaceful come to my mind when I think about nature. Nature is powerful, not only because of earthquakes and things but because of the effect it has on people. People relax in Nature and enjoy themselves. That's powerful in my mind. My feelings about Nature include religious feelings too. Sometimes when I think about nature, I also think about God. These are my first thoughts when you say the word nature. Because Nature comes from God, we have an obligation to take care of it. But Nature has been exploited by many people that do not care about it. Many things have been ruined. Our earth is in trouble. It is being taken advantage of by people that are using parts of Nature that don't really belong to them. This has caused a lot of pollution and wasting of endangered species. Most of the problems that Nature has are caused by people not caring. Nature needs our conservation efforts. We should be using only the resources that are necessities so that there is something left for the future. If we all worked together we could do some restoration but I don't think that that will happen. People just don't care. Nature can be understood although it is very complex and sometimes difficult to understand. There is an order to part of nature. Things like food webs or plant life cycles can be understood and predicted. There are many things in Nature that we understand now and we will understand more as we go along. Science often leads to understanding interesting questions. It can be used to help in conservation. Scientists and environmental organizations are concerned about conservation and our resources. Other things about Nature aren't so easy to understand. Earthquakes and volcanoes can't really be predicted and that makes them pretty dangerous. The danger there makes these things mysterious to me. That brings me back to my original feelings when I started to think about nature. It's mysterious, I like to think about it. The pleasure I get from being in Nature is very important to me. I spend a lot of time in nature. I'd be pretty bored if I didn't have it. It's sacred.

**About Samantha:**

Samantha is a quiet, shy and thoughtful Anglo-American girl who works hard in school for the grades that she earns. Both of her parents are teachers. Her mother teaches preschool and her father is a secondary English teacher. Samantha takes her education seriously. She plans to go first to community college after her high school graduation and than continue for two years at a state university. The only extra-curricular activity that Samantha has participated in at school is the Drama Club. She has worked on several different plays doing sets and costumes but not acting. In her free time away from school Samantha has worked as a gymnastics coach and enjoys working with children. She wants to pursue a profession in mass communications or journalism. Although, Samantha has not been very successful in the science classes that she has taken, she likes science and is very much concerned about environmental issues. She has spent considerable time outdoors with her family and the aesthetic pleasure that she derives from Nature is evident when you talk to her. Samantha has a GPA of 1.81 with a class rank of 109/144.

**Ann**

Nature is something that is always out there and it will always be out there. Everything that exists is a part of Nature including you and me. To me, Nature is beautiful and pure because it is God's creation. Nature provides both aesthetic and emotional pleasure and I need it for self-renewal. I like to go where you can't see any influence by man. When I'm out in Nature, I feel calm and peaceful. It is a spiritual feeling and it helps me understand myself. I also get a spiritual feeling from nature. Sometimes, when I'm out in Nature and I have time to think, I start to wonder about things. This leads me to ask questions that I'd like to find answers to. The pleasure I get from Nature is enhanced by the mysteries I see in it. I also think about how caring about nature. Because it is God's creation, we are obligated to take care of it. Nature's resources are necessities but they are also limited. It bothers me that people are so greedy and use nature. They take things for granted and don't think about the effect that they are having on the world. Many things are now polluted and our rain forests are endangered due to lack of caring. If everybody would learn to love Nature they would take better care of it. Everybody can do little things that would help. I do them every day. If we all had caring attitudes Nature would be restorable. Nature is knowable but the questions I ask about Nature make me think that Nature is sometimes very confusing. It is also changeable. There are some things like the weather that we can predict but other things are not predictable. Some things, like earthquakes, can be dangerous because of their unpredictability. We can learn to understand many things about Nature through personal experience, school and science. Science, itself, provides us with technology, which, in turn increases our scientific knowledge. Technology helps provide us with many wants, which, of course, increases our pleasure. It also uses resources. I know that we can learn about Nature and use that knowledge to change some things that we have done that are bad and to predict and/or control some of the problems that Nature causes us. Nature has always provided for us. We can use our understanding of and caring about Nature to correct some of the damage that we've done and protect Nature for the future.

**About Ann:**

Ann is Anglo-American. Her family has had problems for several years. Her parents have divorced, remarried and divorced again. During the last marriage, they had a second child and Ann has spent considerable time raising him. She is a very responsible young woman, very personable, has a few close friends, and is a concerned student. Ann's father is the PE teacher at the school that Ann attends as well as the football coach. Ann, however, shows little interest in sports, although she is very close to her father. Her mother is a health care worker and currently attending college part time to earn her nursing degree. Ann is very proud of her mother's educational efforts. Ann's family supports her strongly but they do few activities together. Ann is a successful student. She has taken two science classes and plans to complete two or three more before graduating. She plans to go to either community college or a state university and would like, eventually, to enter a medical profession. Ann loves being at school and spends a lot of time there. She participates in SADD, the Drama Club, and a very active group called H.O.P.E. – a science club that does water tests for the Forest Service. She also baby-sits her little brother for her father while he coaches or has football games. Ann can be counted on to help whenever asked. Ann has a GPA of 3.86 and with a class rank of 15/144.

**Kevin**

If we lose Nature then we have lost ourselves, so it better be sacred to us or else we are in trouble. Nature is everything around us that was not made by man. Nature is central to our existence. It provides us with both beauty and resources. I live in a natural area. Being in Nature is important to me. I can see and feel it so I know it exists. I enjoy the beauties of nature, the animals, mountains etc. It supports my sense of self. I believe there is a purpose behind Nature but I don't know what it is. I'm not sure if there is a God or how God and Nature are related. I think Nature is very complex. There are unknown parts of Nature and they are confusing to me because there are no real laws controlling them. There is no order. These parts of Nature can be very powerful, dangerous and unpredictable. Earthquakes are an example, also rattlesnakes. I think that because Nature is so important to us we need to work to learn more about it. Knowing about Nature makes us feel more at home in it. There are also knowable parts of nature. We can learn about Nature through science. There is order to some things and we can base predictions on that. Examples of knowable, predictable things would be states of matter, life cycles, the earth's plates and sometimes the weather. Nature has always been here, but it has changed due to natural and human influences. The resources in our environment are a necessity to us for our survival. But our growing need has lead to exploitation due to people's lack of caring. Over population, oil spills, air pollution and ozone depletion are a result. I want to be a scientist. Science raises many questions about nature. By trying to answer those questions maybe we can learn to restore some of the changed, damaged parts of nature. Our environment needs protection for the future. We need to protect the environment by recycling, car - pooling, reducing pollution, conserving trees, etc. The ability to protect requires knowledge.

**About Kevin:**

Kevin comes from a strong, supportive Anglo-American family. His father is a general contractor and vice-president of a construction company. His mother is a health aide at an elementary school. When asked what his family does together he responded with watch TV together, have dinner discussions, attend church and take part in family athletic events. He also talks about family travels and outdoor experiences in conversation. He likes to talk about his family. Kevin is a strong student. On a scale of 1 to 10, he rates himself as a 9½ but his teachers would rate him a 10. His study skills are well above average and he actively participates in classroom learning. He has completed three science classes as a sophomore in high school and plans to take three more before he graduates. Not completely sure of his professional goals, Kevin expresses interest in the fields of architecture and design and aerospace engineering. Kevin is very active in extracurricular activities. They include cross-country running, Spanish club, and H.O.P.E– a science club that does water tests for the Forest Service. He also likes soccer, running, hiking/mountain climbing and "women". Kevin has a GPA of 3.86 and a class ranking of 16/144.

**Betty**

I think Nature is basically everything. Man is part of nature. Even though we do some things that might seem unnatural, they are natural because we are natural. Nature is both knowable and mysterious. I think there are a lot of conflicts to be considered when you talk about nature. Nature is knowable. But people know or understand Nature in two very different ways. Some understand Nature on a religious or spiritual level. They know Nature as an emotionally uplifting experience. God and Nature are intermingled in New Age spirituality. Nature has aspects that can be considered not only to be living and but to also have consciousness. Other people know Nature on a scientific or factual basis. Their knowledge is based on facts and can be applied to solving problems as it is logical. There is an order to Nature, which we can use to predict some things, weather for example. Ideas about evolution, the ice age, extinction's and global warning can be developed and studied with scientific methods and proofs. Medical cures are another benefit we've gained through factual knowledge. We can also use our knowledge to make changes to nature. Some of the changes that we have made in the past and are making now aren't very beneficial to nature. We have exploited our resources and caused destruction. But our knowledge has also allowed us to do some restoration. My understanding of Nature is more scientific and logical than spiritual but there are some aspects of both attitudes in my thinking. Nature is complex and therefore mysterious. We don't understand a lot of things in Nature because of its unpredictability. Tornadoes and earthquakes are unpredictable and there are many questions that are still unanswered. Another reason that Nature is mysterious is that it is living. Things in Nature have a consciousness. That is something people don't realize or understand. The consciousness and the beauty of Nature are powerful forces. They affect the way that people look at things and how they react. Nature is not going to last forever. It needs protection and conservation. People should be more concerned about nature. They don't seem to care enough about the future and what will be left for their children. The damage that people do to Nature worries me. Things like pollution and waste of resources are going to lead to more and more problems. People should realize that Nature is a valuable resource. As such, it will be used up eventually. The more careful we are the longer it will last.

**About Betty:**

Betty has a GPA of 2.5 with a class rank of 89/157, but she is currently on homebound instruction. She is being treated for severe depression and related medical problems. Betty comes from a divorced home and frequently speaks of the dichotomy of her parents' interests and personalities. Her father is Japanese-American and works in computer science. Her mother, who is Anglo-American, researches and writes family histories for people. Her mother is artistic and involved in what Betty refers to as "New-Age Christianity". Betty lives with her mother but talked to friends about her desire to either see her parents reunited or at least to have more contact with her father. Prior to leaving school for homebound instruction, she spent much of her out of class time in the school library. She helped out there when she could and she called it a "safe haven" from the stresses of dealing with her peers. According to her teachers, Betty is a bright, intelligent girl but unable to organize herself at this time to be successful in school.

**Paula**

Nature is mysterious. I wonder about nature. I would enjoy living in the mountains where the ground has been untouched by humans so I could appreciate the beauty and purity of the natural world. I believe that there are two aspects of nature; the natural world and the human world. The natural world was only in existence before the dawning of man. God created the natural, which makes it very mysterious and, for the most part, is unexplainable. God intended it to be here for a purpose, which is only known by Him. Because it is God's, humans have no right to mess with it. Even with the best technology and scientists we will probably not every fully understand nature. When man entered this planet, he destroyed its purity, beauty, and power. With the exception of hippies, who value the spiritual ideals, the emotional values, and the mystery of nature, man has doomed the planet. I don't understand the human world and why people feel the need to study nature. Studying Nature only causes trouble. It creates more technology and curiosity, which leads to the exploitation of the land. We are stripping the natural world of all its raw materials such as water, minerals, and plants vital to the Earth's survival. The overuse of these materials will doom us, not to mention the buildings, clothes, and machines that make the natural world unnatural and polluted. Our society has the hippies and activists to save the world, but there are so many unnatural things being produced, like growing industries and the production of cars, that make the precious natural world a part of history, never to return. The natural world without human interference is self-sufficient and self-sustaining. The mysterious natural world can sometimes be unpredictable which actually makes it interesting to think about. People must learn to live differently if they want to keep this place. It is a very spiritual world if man's technology would not interfere with it.

**About Paula:**

Paula comes from a troubled Anglo-American family. She is an extremely stressed student due to her personal and social life problems. These problems cause her to frequently miss school, thus, affecting her grades. When attending her classes she is an intelligent and creative person, making good grades and participating in class discussions with vibrancy. She is not self-disciplined when it comes to her studies, however. Paula is called a flighty or spacey "new age want-a-be" by teachers and friends. She seems to be a very progressive, spiritual person who would probably fit in well in a Sedona[40] type place. Unfortunately, because of her lack of discipline she dropped out of high school at 14 years old. At that time, Paula had no GPA (zero) and a class rank of 140/153.

**Patricia**

God created the Natural World. It has many characteristics: it's powerful, diverse, changeable, and beautiful (pause) physically and emotionally. Nature or the Natural World is anything made by God, all the plants and animals on earth and the entire solar system. The Natural World is very mysterious to me, I wonder about many things in nature. Something I wonder about is, what is way out in the universe, perhaps another earth? Even though Nature is mysterious, everything is knowable but maybe not in the near future. The wonderment of the world increases knowledge through science but is limited due to its complexity. The Natural World has many different aspects adding to the complexity of it. It's always changing, the same thing doesn't happen every day; an example would be the weather. Because of change, it can also be beautiful in a naturalistic way. You don't have to know about things in Nature to recognize that they are beautiful and sometimes pure. Some aspects of the natural world are understandable. Science provides ways for us to use resources but also ultimately exploits those resources. In terms of religion, the Natural World is knowable because we have faith in the purpose of it, even though we don't necessarily know it. There is some conflict between the Bible's teachings and views of scientists and environmentalists. Both worlds, science and religion, try to explain the hard questions such as the origin of life, in which I believe there is no true answer. Science and religion have distinct roles in our life teachings. Science teaches us how to conserve our resources and how to possibly restore them. Religion teaches us the caring attitudes required to be productive members of the natural world. The Natural World also provides us with many resources such as: food, fuel, minerals, and plants that give us cures for disease. Our knowledge of the natural world throughout science allows us to use our natural resources and at the same time exploit them. This exploitation will eventually put an end to Earth life as we know it, if we don't start changing our way of living. The natural world was created by God so we can serve him and care for it. We have taken advantage of it long enough. People must learn to take the time to enjoy the beauty of Nature both religiously and scientifically.

**About Patricia:**

Patricia is Anglo-American. She is a religious person and believes the natural world is the work of God. She comes from a fairly close-knit family that enjoys going to church, football games, and soccer games together. She has one older brother who she looks up to. Her father is a police officer and her mother is a bank supervisor. Patricia is an excellent student who has leadership skills and has an open-mind about new ideas. However, in the classroom she is easily distracted. Her special

[40] A city widely known as a Mecca for New Age religionists.

interests are: student council, track, church and volunteer work with the "make a wish foundation." She plans to attend a state university and to eventually become a pediatrician. Patricia has a GPA of 3.60 and ranks 29/153 in her class.

**Jackie**

Nature is alive. It is everything around us like plants and animals. It does not include far off planets because we don't need to deal with the problems there; we have our own problems. Everyone exploits the natural world. They poach endangered animals, destroy habitats, and strip the earth of our important resources. These resources are essential for life and exploited them not only leads to depletion, but using them leads to pollution that is destroying our ozone layer. Some people restore nature, but because of lack of care and laws regulating technology, it is not being done properly. Everyone needs to recycle, carpool and even use alternate means of transportation for us to restore the natural world. There are so many things to think about when dealing with nature. It is very mysterious; how did the earth originate, how does a baby form in the mother's womb, which makes it very confusing and unpredictable. The lack of predictableness, can be dangerous, such as hurricanes and earthquakes. I don't mean to say that it is totally mysterious. There are some things that can be knowable, like medical science. We learn more every day through medicine like how to prolong life. Knowledge of the natural world also gives us the information needed to restore our world, such as the use of electric cars. We get some of this knowledge by taking science classes and through the media, like the discovery channel. We can learn about how technology is providing us with better and more efficient modes of transportation, and developments of solar power. The media is used primarily to warn us about the side effects of using products with CFC's and other damaging things that we are doing to the natural world. Our natural world is endangered and people must learn how to take care of it. The solution is recycle. The Natural World is everything that God created and therefore has a purpose. The Indian culture believes that this land is sacred and should be taken care of.

**About Jackie:**

Jackie is a very reserved person reluctant to allow people to know her personally. She is Native American and sometimes seems to be ashamed of it. This is not to say she is ashamed of her family. She holds her family to be of great importance and enjoys being with them. Anytime there is a "happening" or sickness in the family, she goes to Winslow[41] to be close to the people she loves. Some of the family's interests are camping, going to movies, and riding motorcycles. She lives with her mother and father. Her mother is a waitress and her father is an engineer for the Santa Fe Railway. Jackie enjoys sports and Spanish class. She plans on going to college or to enlist in the military to eventually become a police officer or a physical therapist. Jackie has a GPA of 2.90 and a class ranking of 68/153.

---

[41] Winslow is a city that has a high Native American population and close to Indian lands.

**Bruce**

The natural world is a world that is pure, unpolluted- a place where everything can live together. Nature is complex, but it is orderly and knowable within a 20% or so margin of error. Most of the natural world can be known through science and the theories that have been developed by science. Science enables us to predict, to some extent, everything such as weather, volcanoes, earthquakes and earth movement. Our knowledge is limited by hard questions, such as why does the earth spin the way it does, what is gravity, and why is our planet solid and not gassy. This mystery and the knowledge we have, leads us to a sort of philosophical sort of beauty. Having an open mind allows you to see the beautiful things in nature, like life in the Sahara Desert. Some people think it is ugly and a wasteland, but if you think about it, it has an abundant amount of life, which makes it beautiful. It is beautiful in different ways. There is the physical beauty and there is the emotional or amazing beauty. An example of amazing beauty is the way you think like when you look at the Sahara Desert, there is life in a seemingly dead place. That is incredible to think about. Questions about how the natural world works is explainable through science. Chemists and astrophysicists can come up with theories based on the order and the predictableness of the phenomena in the world. Some things scientists know about are: weather patterns, El Nino, Ozone depletion, and tectonic plate movements. The natural world is exploited because of us, humans. The earth is in danger because humans are destroying the ozone layer, rain forests and precious land. An incredible things is that our resources are being exploited and used up, and we need these things to continue our life on this planet. Resources are the key to survival and we are destroying them. We can restore our natural world to its natural state by conserving. We must learn to re-use our garbage and shrink the size of landfills. An incredible thing to think about is that we all live here and we are destroying our biggest resource, The Natural World. The Natural World needs our protection. We must learn conservation techniques in order to protect our resources so to avoid the damaging effects of pollution. If we act now by not burning the rain forests and stop dumping hazardous wastes we can avoid doom and the endangerment of any more animals. As humans, we have personal and religious obligations to our world to take care of it.

**About Bruce:**

Bruce is Anglo-American and comes from a strict Mormon family. It seems that because of this, the purity of the world, family structure, and success in education are very important to him. However, he lacks maturity and critical thinking skills. He often does not get the course grades he desires. His mother plays an important role in his life. She is a middle school teacher. She understands that science is a critical class in high school and does everything she can to help him achieve at a desirable level. Bruce wants to attend college at BYU and major in chemistry in order to become a dentist or a chef. He categorizes himself as being a very serious student although he slacks off a lot. His family includes three older brothers (two of whom are at BYU), his mother, and his father who is a dentist. As a family, they enjoy going on skiing trips and other outings. They are a close-knit family that influences his life greatly. Bruce has a GPA of 2.77 and a class ranking of 81/157.

**Sally**

I think of the natural world as what God gave us to take care of. In the Bible, it says we are superior to animals and plants. So, we are supposed to take care of them. Religion teaches the caring attitude people must have in order to conserve our natural resources. We have an obligation to take care of this world because God created it for a purpose. We don't know this purpose because it is beyond mortal thought. Knowing that the natural world was given to us by God gives me a wonderful and uplifting feeling. God intended the humans to be the superior-powerful people they are, not so they could exploit Nature but so they could become stewards of our Earth. It is a beautiful place, not only physically but emotionally. The natural world is somewhat knowable through science and religion. It is too big to be entirely explained, for example, how can you be sure that an animal is truly extinct if you can't explore all areas of the world. Science and scientists help us to know some of the natural world because things can be predicted, like animal behavior. The predictableness allows us to answer how things work, but we will never really know why things work: Why is Nature here, What is the purpose, or, How did life form? Some things are unpredictable like hurricanes, tornadoes, and volcanoes, which make Nature dangerous at times. Science can teach us how to be better conservationists through research and technology so we can avoid pollution which ruins nature. Sometimes people have too much power. Some of the uncaring with no religious background exploit Nature by developing nuclear bombs, destroying land, ruining our rainforests, and endangering animals which will possibly doom us. Regardless of these people Nature will survive because of the many people who do care. People shouldn't have the power to destroy. They should only have the caring power. God would not have given all this beauty for us to ruin. We are stewards of God's land.

**About Sally:**

Sally is Anglo-American. She is a religious person who believes the natural world is everything under the influence of God. She is a very serious and sober person who knows the value of an education. She also has a serious attitude about her special interests, which are drama, honors choir, and Falcon fundraisers. She plans on attending college and pursuing a career in marine biology, child psychology, or public relations. Her mother is a bartender and her father is a cable repair employee. They enjoy family holidays, going to movies and football games together. Sally has a GPA of 3.79 and a class rank of 24/157.

**Allen**

The natural world is not man-made, it is plain living. It is everything that man does not interfere with, like outer space and the stars. Nature is knowable to some extent; like people can recycle and fix the ozone layer by not driving cars and stuff. If all people would care about the natural world, we could restore it so it would not be doomed. People need to realize that they are killing animals, plants and eventually themselves by polluting our earth and not doing anything about it. Nature gives us all we need to survive, like food, water and air along with the abundant resources such as coal, oil, gas and wood; so if we don't conserve these, the natural world would not exist as we know it. People know what they are doing when they burn the rain forests, pollute the oceans, and drive cars; why don't they care? There are things we can do to fix the earth so it can be restored to it's original state which would be very peaceful.

**About Allen:**

Allen is a hardworking Anglo-American student. He wants to be a good student but his learning disabilities hinder him. He is a "resource student" that tires very quickly with his studies. He only functions while in school and really does not have the self-discipline to do homework or study for tests at home. He does not participate in any extra-curricular activities and does not enjoy staying around at school. Away from school, he loves to race dirt bikes. He would like to go to college and major in business but would really like to race dirt bikes professionally. He is a very caring and sincere person. He has much family support from both of his parents, who own a car dealership. Allen has a GPA of 2.20 and a class ranking of 117/157.

**Holly**

The natural world is just there, you know, fish, bugs, dirt, animals, and plants. There are aspects of Nature that have purpose because it was probably created by God, but I am not really religious so I can't explain it. It is very big and complex, like the ocean, which makes it somewhat confusing to know about. There is some order in nature, but not much. An example would be like some parts of the land, like deserts, forests and oceans. Humans are included in nature, but not the things they make, such as cities. Cities are not included in the natural world because they are built by unnatural means. Because of these cities, the natural world is exploited; like our resources that we use for medicines, paper and breathing. People need to realize that our resources need to be protected because they are necessities for life. They can be recycled. I do not recycle because it is probably not in danger now or during my lifetime, so what's the point? These resources are abundant but not unlimited. The natural world is being ruined and people need to begin restoring it. I myself really can't restore nature, but everything together can.

**About Holly:**

Holly is Anglo-American. She comes from a divorced family and has an older brother. She does not have a good relationship with her family and would rather not be at home. Holly shows little concern for much of anything. She does not seem to care about much of anything. Holly is a reluctant participant at school. She always seems reluctant to give any information or to discuss her own thoughts and ideas about anything. This makes it difficult to have any kind of conversation with her. She is not a good student and sometimes refuses to participate in class discussions. She enjoys talking on the phone and "hanging out" with friends. She dresses in a very rebellious fashion and enjoys looking and acting differently from others just for the reaction it provokes. Holly has a GPA of 2.79 and a class ranking of 78/157.

**Liz**

The natural world is all the animals and the things around us. It also includes the environment and how they interact with it. The natural world also consists of ideas, why animals do certain things, their purpose, and what they think. It is the work of God. Its purpose is to help us live and enjoy the things - the beauty it provides us. Everything happens for a purpose. The natural influences the way we think. Just thinking about the natural world gives us peace and energy by knowing that the animals are okay and to see that we are not the only organisms living here. I live out in the desert where I can enjoy looking and thinking about the animals that live there. I get peace from that. It also provides us with things to think about like, what's going to happen to us if we don't learn how to protect this earth. The natural world is knowable by means of education through science and by learning through personal experiences. Eventually we will probably be able to know most things about the natural world. However, some things will be kept a mystery because not all things are meant to be known. Science tends to teach the how and what questions about the natural world and religion hints at the why questions somewhat. Before it can be knowable to someone, that person must care about the natural world. Lack of care not only hinders your personal thought but sometimes leads to exploitation of natural resources and natural environments like the rain forests. Also, people like hunters take advantage of it by killing animals just for their skin and the fun of it. That is a waste of the natural world. Because the natural world is knowable, it allows us to restore our natural resources by conserving them and uses of technology to possibly find new ones. However, it is difficult to understand why people don't care enough to save our earth and not destroy things. If we are going to use technology and the natural world to our advantage we must first learn to care so we can put an end to pollution which could eventually make us doomed. The natural world is here for a purpose, and that is to help us in our studies and for us to use it rationally. We must be careful because this is the only earth we have.

**About Liz:**

Liz is Anglo-American. Liz believes that the natural world is everything around us and helps "decorate the things around us". The latter part of the definition is indicative of her personality. She is a very colorful and vibrant person, always smiling and happy. Liz is a person who has no enemies due to her positive outlook on life. This positive attitude is consistent in her family. Her family does many things together. They go to movies and sporting events. They take family vacations. They like to ride horses. Liz has an older sister who she looks up to and shares her thoughts with. Her dad is a builder and her mother is a housewife. She plans on attending college on a softball scholarship and pursuing a law degree. She is a very outspoken person and an active participant in all of her classes. Liz has a GPA of 3.71 and a class ranking of 27/157.

# *Appendix B*

Teacher Narratives on Nature

**Mr. Bradford**

Nature is the living and non living components of the world around us - even the universe - apart from the works of man but including man. Nature has been here forever and it will always be here whether man is here or not. We are all interconnected in Nature by both natural and artificial mechanisms. We, as a species, are all part of the same natural world and we all have an effect on each of the other components and species of the natural world. Nature in its natural state is pure and perfect. It can improve itself but it cannot be improved by the works of man. Purpose in Nature means the struggle for survival. It does not mean there is a god who gives purpose to Nature and directs Nature by will.

Nature is ever-changing. If man weren't there, beavers, for example, would still change Nature! Plants invading the land have changed Nature. There are many different aspects to Nature. The more diverse Nature is the more it is the way it should be. For example, in the ideal forest, in my mind, there is a wide variety of plants and animals, all different kinds of trees. Whenever that ideal forest is altered by man, there seems to be less and less diversity among those types of organisms in Nature. Nature, however, is more powerful than the minds of people who are trying to conquer it. Eventually, Nature will win out and so far it has won out, because man has yet to conquer all of Nature and I think that... because of its complexity, because of its mysteries, because of its unsolvable Nature, it remains to be powerful, more powerful than man. Glenn Canyon Dam is an example. The dam has permanently altered an aspect of Nature. Man conquered that section of that river, but he has not conquered the entire river. Eventually the water will flow over that dam and destroy it, so Nature will be more powerful than man, in the long run.

Due to the diversity in Nature, Nature is very complex. All of the various components of Nature are working together, and in some cases working apart. It leads toward the complexity of Nature. It makes it very hard to figure out. It makes Nature mysterious. Nature is mysterious because it is so complex, the diversity of it makes it mysterious. There is a lot that is not known about Nature. No one will ever know everything there is to know about Nature and that is part of its appeal; because it is so mysterious. Not only will nobody ever know everything there is to know about Nature, hopefully no one ever will.

To me, the mysterious Nature of Nature is one of its better qualities. Things that are completely discovered are no longer interesting. For example, you have a cube of metal that everybody knows every single ingredient in it. Well, there is no mystery to it. There is nothing appealing about that anymore because there are no questions to ask about it. If everybody knows everything there is to know about that cube of metal, it looses appeal to me and I am sure that it looses its appeal to the people investigating it. If things don't have questions associated with them, there is no mystery. If there is nothing to ask about it anymore, it looses its intrigue, its interest. The mysteries of Nature are hopefully unsolvable. I don't want to solve all the mysteries of Nature. I hope nobody ever does. The appeal is like being lost out in the forest, so to speak. You want to be out there away from anything that is solved, you want to be in an environment where everything is still interesting to you.

We know some things about Nature. We know, for example, that in Nature you will find examples where the progression of an organism is a direct result of the purpose of that organism - where the purpose of organisms is just to survive or to carry on their genetic traits through generations. Certain laws of physics can be applied to just about any part of Nature so there is some predictability in Nature. Water *will* evaporate. It *will* rain. You can predict certain population outcomes based on certain criteria like the amount of rainfall. We know these things from science. A scientist makes observations and collects data. He experiments with controlled experiments. He makes observations without experiments. He experiences Nature. He lives in it. There are a number of ways to study Nature. Scientists, in general, probably do more of it than anybody else. I think there are a lot of biologists who have made it their life's work to study Nature. They do it better than anybody else. I think everybody should do experiments, either in their head or scientifically -on paper, in Nature, wherever. It should be done. Does everybody actually do it? No, they don't, and that is part of the problem of why Nature is being destroyed.

It is important, therefore, to understand how things work in Nature because we are effecting Nature all the time. For example, we effect one thing and it effects several other things that will have an effect on me or other parts of Nature. This indirectly affects me again. It is important to me to have people understand how Nature works so that they can prevent effecting it more than they do. Nature is difficult to understand - remember, it is very diverse. People think that they understand Nature and so they go ahead and affect it the way that they want to. They predict that they won't affect anything else, but in fact, they do. So, I think that it is important to understand how Nature works, as best we can, so as not to destroy it. I think everybody should study Nature, I am not sure that everybody does do it. Scientists probably do more of it than anybody else.

Nature, however, is really not knowable. If Nature were knowable it would mean that you would be able to predict anything about Nature. You would know all the components of Nature, what effects it has on other components. Well, you can predict a certain amount of Natures' outcome, but overall, Nature is unpredictable. If we destroy this insect, for example, because it is hurting some tree or some plant that we care about, I don't think that we will ever be able to predict how we have effected Nature as a whole.

You can't predict what effect one aspect of Nature will have on all of the other components of Nature. So, it is unpredictable in that regard.

In my mind the work of god is pure. It is perfect and it can't be improved by man in anyway. It can be improved by the works of Nature itself. God-like things are perfect. They can't be improved by works of man, but can be improved by the works of Nature itself. Nature is pure and diverse, that is, *unaltered* Nature. Pure Nature is Nature unaltered, and so the more diverse Nature is, the more unaltered it is. Unfortunately, there are a lot of cases where that isn't true, but the ideal Nature should be pure and diverse. Because man has shown that he can change certain components of Nature, I am concerned about pollution and the damage that it does to Nature. Right now there is overwhelming damage being done to Nature - the effects of man on Nature in our lifetime are pollution, destruction of rainforest, the damming of rivers, the dredging of the oceans, and the pollution of our oceans, the killing of species, and so on.

Nature is beautiful, as I see it. That is what draws me to Nature in the first place, how beautiful it is. The simple beauty of being pure, the kind of plants and animals that are out there, the landscape in its natural state, all kinds of simple beauty to it. Purity and diversity have an internalized beauty to me. When it is pure and when there is great diversity out there, then it is more beautiful to me. So, those things have to be in place first, possibly before I consider it to be beautiful. Nature is living. Nature is composed of living things organisms, and the living part of Nature is probably what attracts me to Nature in the first place. So the living part of Nature is what appeals to me - plants and animals, any kind of plants or animals. Even though I would consider rocks and volcano's a part of Nature, the living part of it appeals to me more. There are some beautiful rock formations and so on, but the living parts of the landscape is what is most beautiful. The appeal for me is an internal sense of peacefulness when I am around Nature.

**About Mr. Bradford**

Mr. Bradford is Anglo-American. He is a second-year biology teacher with a college major in biology followed by a secondary science teacher education program.. He is a serious person who speaks quite philosophically. It was not clear what his religious inclinations might be. He was reluctant to talk about religion. He expressed strong feelings about Nature and in rather intellectual terms of solving a mystery. Mr. Bradford said little about aesthetics and nothing about resources in Nature, but he clearly enjoyed personal contact with Nature such as hiking and camping.

**Ms Jackson**

Nature is the living world that we live in and it is more of the man-made kind of things based on what we understand about the laws of Nature. I think that Nature is predictable. I think that it is logical. I think that it is explainable. As scientists, we come up with laws of Nature or theories of Nature to be able to predict behaviors and therefore, based on what we know, and the experiments that we have done, we can now, either change or know that we can't change an event, but that maybe we can predict that the event is going to happen. And we can predict those things because they are orderly, there are certain patterns that we can find, and yet at times they can be very complex. But I think Nature, you can understand it, you can know it, and you can predict it. I think that if we study it that Nature is not difficult to understand. For instance, I am not a real biology type person, but I like watching those shows and they show patterns of things having these five sides, so I guess that if you are to get a new plant, then you could categorize it, because of those sides, but basically, from the physical science side, a lot of, if you are going to... gravity is, and you throw a ball, then you can predict what is going to happen because it is logical... I think that is what I'm thinking of when I think orderly... I think logical. If you were going to use lenses, and you know how light is going to go into the lenses, then let's say, with a telescope, you know that you need to use a lens to get the image and then the lens to invert the image, and so how you know how to use it, because of what it does. It is very logical to me.

There are limits to knowledge, however. We don't know a lot about genetics. That seems to be real prevalent today. What they are trying to find out, like pre-birth testing. How can we find out if you are going to have a disease or even if you are alive, are you going to be predisposed to having cancer or to having diseases... so, we don't know everything, but we have enough information to keep trying. And I think that we have the space shuttles that go out, we have the satellites that go out and we try to learn more about space, but there is still a lot that we don't know, but based on what we do know, we have a direction that we can go in. The future is an interesting question. I am a fan of Star Trek. It would be nice to know that we could have space travel and that we could actually achieve that, but I think, I don't know, I feel like we know and awful lot. I feel like, that, even though we don't have all the answers, we have so much, ways of finding out answers. I think we have that base of knowledge, so I think, I feel that we know an awful lot. I think we would definitely be up towards the 80 percentile of knowledge. This knowledge, I think, has come from a variety of sources. I think that there have always been scientists studying things and that has been formally or informally, and I think that, for instance, on this sort of thing, like how to reuse energy, or how to get restorable energy, I think that's coming from scientists, but it is also coming from business, because they need to find a way. So, they're a source. They are using scientists, they are using people that are engineers, along those lines, but they're the ones that are promoting as well as the government. So, I think that the sources are coming from, and I think that people may be in their own houses, if they can come up with something that works a little better. So, I think that it is kind of a variety, but I think that science has a lot to do with that.

We use the resources of Nature. We use trees. We use coal. We use oil. We eat plants. This is why we study Nature, because... that we do use it for so many things... resources each day. It's the things that we use. It's the things that we interact with. Without the things of Nature we cannot enjoy the lifestyle we have today. I think that a lot of things that we make... I don't know all about what materials things are made of, but I know that a lot of things come from... we have leather, we have wool, so I think that because we know and understand the laws of Nature, we can now build all of this electronic equipment that we use. This is more for the laws of Nature, like radio waves, TV waves. That is why it is important to understand how things work in Nature... so that if you want to grow something, that you understand how to grow that better. If you want to use any laws of Nature, for instance how to make... a lot of these are man-made materials, but then you are using laws of Nature, of how to get solid-state equipment and get the electrons to move, or get the radio signals to go. So, Nature is something that should be studied so that we can learn more about it. This is what scientists do. Lots of scientists are doing that. I think little kids do it, and I don't know if they really add to our understanding, but I think that it is a natural thing for kids to do to just start looking at the world around you, and then taking things apart at your house, and finding out how they work. So, I think that it starts with kids... and everybody has a little bit of it, but then some people make it their formal career like scientists.

We need to be aware of what is around us and how we fit into the whole thing. I am a scientist, for one thing, and with so much new knowledge, with so much resources, with genetic engineering that is going on, I think that we need to be better equipped citizens to be better to make decisions on whether or not this is something that we want to do or not do, and if we don't know, where we, as mankind, fit into this... then what is the good of making better people, more people, if there are no resources for all of these people? I mean everyone wants to live longer, but there is a purpose for dying. So, that sort of thing... so that we have new people! You know what I'm saying... it's just that we need to see where we fit into all of this, and we can use all of the laws of physics to go travel everywhere that we want to go, but why would we want to do that? Do we want to have more people so that we can live other places, you know... do we have room for them... but where do the resources go? I don't know if this is true or not, but somebody just told me yesterday that when you used to be able to eat one apple and get all of the minerals you need, now you have to eat five. I don't know if this is true or not. It was an interesting comment, so it made me think. Their theory was that there is not as much nutrients in the soil. So, this goes into the fact that we're (we are all part of Nature) endangered. We're full of resources. We're exploiting it. We're not either using what we have properly. We're over-using other things and obviously pollution is a problem. Resources are exploited. Nature is polluted and endangered because of those reasons.

But we have all those resources, this material (or matter which is just what, everything that has mass is), and hopefully we can restore it and then we'll have all this matter that will matter to us! I just went to an energy seminar about restorable energy sources, and they are talking about trees, and how we could plant trees, and with the science that we know now, how we can make trees grow really fast or really tall, and how we can use more trees. I think that Nature is reliable. I think that you can count on having what we have, at least for in the short term. That sounds like… and I do, I have hope... and I guess that I have hope that there is… I think that there is a large awareness of what is happening with Nature. So, I don't have this fatalistic kind of attitude about Nature being doomed. I don't know if it is human Nature, or not, or optimism, or... I don't think that I am a necessarily optimistic person. I try to be realistic, but I'm not... I don't know... I would hate to feel doomed, to think that down the road, somewhere, that all is going to blow up! I think that people... I can't help but feel that people will try and preserve Nature, but I don't know that Nature will always be as it is now. I don't know what to do. I don't think in my lifetime… I think in my lifetime that it will be the way it is. But I also think that I don't know enough as a citizen, exactly what is going on. I know that there is a concern about ground water, for example, and whether it is being regenerated enough for us. I don't really know what the studies there are about our increasing population and what we are using of water, and how exactly we are restoring it. Nobody is really yelling real loud about it, so that gives me hope. We would be in trouble if our water was polluted. We would have severe problems. If we were running a plant that, not necessarily a nuclear plant, but something that had highly explosive chemicals, and it blew up, it would hurt people. So you need to know that this is not only pollution, but anything that you are dealing with that could be harmful to Nature. Hopefully by knowing about Nature, you are protecting it and you know what it needs to thrive in the living kind of sense. In the technical kind of sense, the laws of Nature, you are not abusing it, so that you end up destroying what it is that you are trying to promote, which is hopefully life on earth. So, I think that people like to know things about Nature. The scientists come up with stuff and then later on somebody finds a use for it that could be harmful, but that is why you need to protect it. I think that people are always going to know dangerous things. So, I am hoping that there is a balance between use of resources and protection just because this awareness of Nature keeps being raised, and people keep coming up with solutions.

I think because we live in the world, we have to appreciate it and I think that most people do. We all appreciate... I've lived in a variety of places. I have lived around a lot of mountains, around the Alps and I've lived in the Blue Ridge and Appalachian mountains of Virginia, and it has taken me a while to appreciate the beauty of Arizona... just in a natural setting, and I think Nature is beautiful. I think about Nature everyday in one way or another. If it's not the laws of Nature, driving with my kids and I am pointing out the moon to them in Arizona, and like I've said that I have lived a lot of different places, and the sunsets here are the most beautiful sunsets, and I know why we see those sunsets, but it is just nice to enjoy them. I also think that science is beautiful in the fact that you can repeat patterns and that you can find these things that are logical and I just like that. That appeals to me. Because of the physics and the refraction of light you can understand a beautiful sunset.

But I think that Nature itself is pure. I think that if there is nobody that is interfering with it, that it seems like it takes care of it's self. There is a cycle that it goes through. I think that it can be very peaceful, on one side, and of course, like that one picture of a volcano erupting, it can be not so peaceful. But, I think that it is sacred and holy, just because we should take care of it, and we should respect it. Nature brings out emotions and I think that because of that we should respect it, definitely from a religious side as well. And I think that, you know, and that goes for our own bodies... everything. That's because that's more of a religious side, that we should take care of everything. I think that we should protect Nature... you know there is that saying that we are just borrowing the earth from our children, they're not giving it to us

There are other aspects of Nature. I think that there is a lot of living things in Nature. Not everything about Nature is living, especially if you are talking about laws of Nature, that is something a little bit different, but Nature is all centered around living things. I think that it is exciting to study Nature. It is very diverse in terms of, if you look at the chemistry of it or the physics of it or the biology of it or the enjoyment of it, just how those things tie together. Nature is not always peaceful. That is things like earthquakes, things like... and these are things that we can maybe predict but not control, maybe minimize damage. Things like, maybe something falling out of the universe, the sky, like a meteor. Things that are frightening and they're dangerous because they could hurt people. They are powerful enough to have that kind of effect, and I think that the confusion comes from the fact of *why*, especially when someone that you like is hurt. If it is a devastating thing, you just wonder *why* that happened. So, an earthquake is not confusing in the sense that it happened, because you could be able to predict that, but just... Oh gee....you have this disaster and that is sad.

On the other hand, I think things happen in Nature because of purpose... this I think is a religious view... that, not necessarily a fatalistic view, that you have no control, but that there is a destiny, that there is... as an individual, I'm contributing to it, but I'm not the biggest part of it, of human kind that there is. I think there are purposes. Animals have instincts... Humans are different than animals in that we seem to be able to reason and really take control, and again I don't have a real biology background, but I observe those things, or watch shows on them, and so they have these cycles of their life which must have some purpose and it has a purpose for the food chain and how they are all, the whole huge, inter-related processes, so I think that when one... supposedly... I think that I read this, that when one animal gets killed, it's usually the weakest one, and so that they are promoting the stronger one, so that would be a purpose, and that is one of those things that happens in Nature, because of purpose. Now, when man seems to get involved, I don't know that their purpose is to only... we help all our sick people, and then we... through other things... maybe through Nature, but maybe through our own design, then we harm them as well. It's not everywhere that we see that, but I think that it is the way it is meant to be. I think that the whole inter-relatedness of us and our world, is that... like okay we do have a purpose, some bigger picture, and I think that we are playing that part, but I don't know what it is going to lead to. There is no "lead-to" where we are all taking care of each other and we are somehow living in a balance, or are not. I don't know how it will end up. I do know that... I know that all things want to continue living, so that they all reproduce and that seems to be real important in Nature, for plants, people, and animals. I know that we need to use our resources wisely, that is real important, and that is about all I can think of now.

**About Ms Jackson:**

Ms Jackson is Anglo-American. Her college major was physical science education and she teaches physical science courses at the high school. She has several years of teaching experience. She was very positive about the importance of scientific knowledge how much science has taught us about Nature. She spoke about the importance of natural resources and only lightly touched upon exploitation and pollution concerns. She said little about aesthetics or religion. Ms Jackson said little about personal experiences in Nature.

### Mr. Hess

Nature is orderly and understandable. The tides and the rotation of the earth, the seasons and so forth are examples of order in Nature. That the planets and the stars are governed by physical forces and any deviations are simply because we have not yet discovered the other part of Nature's orderliness. According to chaos theory, even things that appear to happen randomly have patterns. I think that everything has patterns. We haven't necessarily discovered those patterns, yet. As a science teacher, I feel that with enough scientific knowledge we all things are understandable. I think it is very important to know how matter interacts with matter, and therefore how that influences everything else around, for example, how living things work, how it rains, how the stars are made, and how they are formed, the whole thing. I think that the more we understand about matter itself, and the more we know about how to make things, the more predictable Nature will be. Scientific or reductionistic thinking is very powerful. I feel that once we know enough about the minutia of the world, breaking it down by using the scientific method, scientists tearing it apart and analyzing the parts of Nature and seeing how they interact, that we will be able to predict just about anything about Nature.

I think there is probably a limit to predictability in Nature. I think Nature has unpredictability because it is so changeable. Weather is a prime example of that. We can't predict the weather more accurately more than, at the most, two days ahead of time. That is because there are so many things that change within this thing that we call weather, that effect local weather patterns. I think that it is probably the best example of something unpredictable. I think that Nature is always changing and that is why we are having a lot of natural disasters. Nature is changeable and we just can't predict that too well - when a hurricane will strike or when a tornado will strike.

I think unpredictability, however, comes because we don't know enough about Nature to predict everything about it right now. This has emotional consequences. We have emotions of fear and peace and I think that fear stems mainly out of the unknown. Man is frightened when he perceives what is going on around him and he doesn't understand it. If events are predicted then there is a very peaceful type of feeling. I don't think that Nature is inherently dangerous or confusing either, because that is man's definition of what is maybe the unknown part of it. What's dangerous about Nature or about the natural world is that we can't predict when things are going to happen, when we are going to die, for example. Eventually, however, all Nature will be explainable.

One of the reasons why we don't yet understand enough about Nature, is because the extreme, complex, and diverse type of systems that are involved with it. But I am an optimist as far as it's understandability, as far as that is concerned. Our current state of being is that there are unpredictable events in Nature. Our ultimate state, the end point, is basically knowing very much. Weight wise, we are probably more tilted toward unpredictableness because I think that we are in the infancy of understanding the world around us. I am optimistic that we will eventually know much more. As knowledge grows, we will change the changeability and the unpredictability of Nature. It all will decrease significantly. I have a great faith in man's ability to understand things and take things apart, to get to the bottom of the solutions and things. I think that with that knowledge and the yearning for knowledge, whatever is the problem, we will basically be able to know and being to predict.

It is important to study Nature for three reasons. First, the mere fact of knowing things about Nature is worthwhile itself. Nature is an everyday part of life and I think about it a lot and how things work and how things interact with each other. Nature is beautiful. I see it most in the way things work so well together. I think that I see beauty in Nature more with living things than with anything else. It is the vastness of things that could go wrong in a living organism, and yet it lives. Nature is made of matter. That matter gives us the resources we need whether it is living resources or material resources. Material resources are the raw materials that we can use to build things or to develop technology. Thus, the second reason to study Nature is that the more we know about Nature, the more we can control it and use it or exploit it. The third thing is the more that we can do that, the better our lives are going to be - and, this is sort of a tribute to Man's intellect. You know, how to use what's here.

I don't think that the natural world will ever be any of these things, endangered, restorable, or doomed. It will never be endangered. It will never be restorable because there is nothing to restore. It can't be doomed because, whatever doomed means, the natural world will exist. Whether man continues to exist or not, it really doesn't matter too much. I think that the natural world will always be there, whatever form it is in. No, Nature doesn't, as an entity, and there is no such thing as Nature as an entity, need protection. It doesn't need protection. That is man's need in life. We need to protect Nature so that Nature can provide us with the materials we need. So, if you put man into the equation, like if the equation says that man needs to be on this planet, then this is what we need to do. If we are not concerned with that, then we shouldn't really worry about what we do with Nature.

I think that it needs to be protected, however, simply because I think that there is enough enjoyment in Nature itself, or different parts of Nature, that the beauty of Nature needs to be protected. I think there is a bigger story, though, about why we need to protect and know about Nature. This is such a bigoted statement, but we need to protect the human race. We need to know enough about the ecosystems, so that we can say, "yes, these animals can become extinct because they are not really important." So, those two sides of myself battle each other because I think that there is a lot of beauty in Nature and I think that it is very enjoyable to have these diverse animals. But, I also think that we also need to be realistic and know that we are not going to be able to protect all of the animals. We need to know what animals are necessary for us to enjoy the same quality of life that we now know.

I also have some other thoughts about Nature that are really completely separate from what I have said so far. These thoughts are extremely important because I think that there is a need in man's life for a purpose. Nature or the natural world is everything. Well, it's the universe, including man, and everything man does, and everything in the universe; but, the natural world is not everything that exists. I think God exists and He is part of the natural world, but at the same time, not part of it. I think that the natural world is a subset of God, and not the other way around. I think that Nature can remind us of the spirituality, our own spirituality... I don't think that Nature has a spiritual quality in itself. I don't think the world around us or the universe has any spiritual qualities. So, God is sort of the wrench in that perfect definition, Nature or the natural world is everything, but a necessary part. It is a necessary wrench because the rest does not work without that. Basically, I am talking about this unknown variable called Man and all his ideas. Divorced from pure science and pure fact there is also something called faith which is what defines sacred and holy and mysterious. Although I think we will eventually understand a great deal about Nature, I also don't think that we can every discount the idea that there will always be a need in human lives, where things are sacred and holy, with holy perceived as mysterious, as well. Even if things aren't mysterious any more, I think that man will invent new things to have as mysterious.

I definitely think that there are parts of everything that are separate from, not Nature, or the natural world, but definitely from what I perceive as what science can uncover, and maybe part of that has to do with man's need and wanting for, and maybe personal discovery of things that are holy and sacred, or mysterious. That is sort of an unknown variable, that sort of sits out there. These aren't products of man's interaction with that part of the natural world. Things become holy because they are a part of the natural world that doesn't fit into a nice little formula, but is somewhere out there that we haven't tapped into. Maybe that is knowable, maybe that is predictable, eventually, but it can't be broken down by using the scientific method.

**About Mr. Hess**

Mr. Hess is Anglo-American. He is a physics teacher who had been teaching for several years. His college major was physics followed by a secondary science teacher education program. He always spoke of science in strong positivistic terms. He saw science as a tool that people use to better their lives. He emphasized the importance that science has played in the discovery and development of natural resources. Mr. Hess used little aesthetic language. He expressed some concern about environmental issues but judged "Man's influence on Nature" as generally positive. He expressed a few vague religious ideas such as giving to "God central position in holding Nature together" but offered nothing more specific than this. He said little about personal experiences in Nature.

**Mr. David**

The natural world is the environment and world around us that is here naturally, without being effected or changed by man. The natural world is what is here that hasn't been changed or influenced by man. I think it is sort of the raw material that we've come upon in our activities. It is sort of altered and changed, but I think the natural world, by calling it natural, you're saying that it is something that has not been altered by man. So when I think of the natural world, I think it includes everything that was here, that we come in contact with, or that we are in contact with. Its all just part of everything that is there. Now that I think about it, I would consider the natural world to include the world, the physical part that we see, and whatever it is that may be behind it, that is created or is driving it - all the parts of it, the parts we understand and the parts that we don't understand. Some might say that they see the work of God in Nature. I would say that you see something beyond the work of man, that's even a higher level.

Nature has many aspects. It is alive and it is always changing. It has a mind of it's own and in some ways things happen, because it is alive. Just the way that the earth moves and shakes, way that the oceans tend to move and the whole relationship between the earth and the universe. The way that living things have come out of all that, or part of it, to interact with the earth and universe. I think that the fact that it's alive really is a big part of what makes it the natural world, or at least my concept of it. I am not using "alive" in the technical living things sense, but I think in terms of how matter (Nature is material as well) interacts. I think that it is alive in the sense that, even though it may not technically be alive, I think that when there is heat and there is energy, things are moving and flying, that in a way is a kind of life. Nature is dynamic... movement and change and all life, when you look down to the molecular level, it really is just non-living, material molecules that are organized in complex ways. So, it is hard to draw the line, when you get to that level, as far as what is alive and what isn't. So, that's partly what makes it mysterious. Nature is alive and it is material.

Nature is orderly and chaotic, predictable and unpredictable - these pairs are sort of needed in order to define each other. Things wouldn't be predictable if you didn't know what unpredictable was. Things wouldn't be orderly if you didn't know what chaotic was. It is sort of a ying-yang relationship between the two I would call this just the dualistic Nature of reality. A storm in the ocean might be considered chaotic, but then as you watch the ripples of the waves that are flowing away from it, there is a sort of orderliness to that. Weather is unpredictable. You can't predict what's going to happen, but you can predict the consequences of it. The resources that Nature contains is kind of unpredictable, because we don't really know what resources are there. The fact that you use the resources of Nature means it is more predictable. But, it is so powerful that we can't really always predict what Nature will do or control it. You can predict that you are going to have certain consequences, however. It is also powerful. In relation to man, Nature is powerful because it controls whether life can exist on this planet or not, or any particular place. And we are real limited in our environments that we are able to occupy and so in that sense, Nature has a lot of power over us.

There is a lot of diversity and complexity in Nature, and there is also the fact that it is just there. It's all just part of everything that is there. You can look at it all as being part of one thing, or you can look at it all as being different and complex in different aspects of it. It is incredibly complicated. The closer you look the more complicated it is and in order for it to function as simply as it appears to us, there must be a lot more to it than we know. I think that it is important to understand that there is more to Nature than meets the eye. It is interesting to see how Nature works and just how complicated it really is. By observation and by looking at things and watching them over a period of time, you begin to notice patterns that allow you to make predictions. But it seems like a lot of predictions, once you make them you find that they... well, the rules tend to get broken, or you get more information at a higher, finer, more detailed level and you realize that there are other things going on that you weren't predicting. I don't see these aspects of Nature as being balanced. I think that it is just different aspects of the same thing and that the pairs both have to be there in order for Nature to be what it is, to define what it is.

I like the word beautiful. I think that there is a lot of beauty in Nature, even though it is not always beautiful to man. The whole aspect of Nature and I guess that I have an instinctual connection to that, that it is sacred, and just deals with something very speciäl, you have to respect it. I think that beauty is the more aesthetic reason to appreciate Nature and I think that aesthetics can provide reasons for studying Nature, too. But, I think that beauty and emotional response are more in the aesthetic realm, just pleasing to see how Nature works, seeing that it is mysterious, that it allows a curiosity about how it works to admire the beauty of Nature and it's simplicity, and just... Well, I enjoy Nature.

Some people might say they see the work of God in Nature, that is to say that you see something beyond the work of man, that's even at a higher level, and to appreciate that is one of the aesthetic things that we like about Nature. I have an instinctual connection to sacredness of Nature. It just deals with something very special about Nature, and you have to respect Nature. As I said before, I think that it is important to understand that there is more to Nature than meets the eye. We need to treat it as being sacred, because of that. I think that Nature is our home and it gives us life, and everything, that people consider it to be very sacred and holy, and those are definitely the perceptions that we have as being a part of Nature, and it kind of grows out of that, I think. These ideas are religious and philosophical and emotional, all three! I think that people have real strong emotional ties to Nature, in a lot of ways. There is a lot of variations on how people consider it to be sacred or holy, but I think everybody does, in some sense. Just about everybody has some connection to that, although I think that when people have self-interest at stake, they tend to ignore those aspects of the natural world, to meet their own needs. Everybody has a sort of innate appreciation for life and for the natural world. I think that a lot of our religious belief has to come from this aspect of Nature, its beauty and mystery and sacredness, which is that part that we can't... or we don't understand at this point, and that it makes us think that there must be something beyond our level of understanding, from where Nature must have come. You wonder about how it all got started and where it all came from? Its very religious, emotional, and philosophical too.

I am look at the natural world as something that is physical, more physical, and it is happening around us, but if there is a God behind it, that is creating it, then that might be something that is at a different level, that I don't understand in my own reach. From what we know about energy and physics and everything, I think that there is potentially other realities or perceptions, or things that are happening, that are beyond this natural world that we are perceiving. I think that there could be things that I just don't know about. I do think Nature is more than material. I think that there is something driving Nature and causing it to evolve the way it has changed, and to say that Nature is only the stuff that you are looking at, is... may be missing part of it. I think that there must be some purpose for things to become what they become, in terms of living organisms, and what is driving it, I don't know. To say that everything is driven by a purpose is, I think, sort of a human perspective. I think that it is an assumption to say that everything results because of a purpose. I think that is possible that things happen because of chance, too. Purpose sort of denies the whole role of fate and chance, that things can happen just because they happen to work out that way! So, I think that you have to have an element of chance in Nature, it is just not purpose. Maybe in terms of any purpose that there might be, there is God, but in terms of being exclusively because of purpose, I just don't think that Nature is playing out according to some divine purpose or that it is all laid out. I think that there is a definite element of chance.

I do think about Nature quite a bit. Just wondering about how things work. When I see a bird fly around, I wonder how its eyes are so quick, how it's wings can move that fast, how quickly it perceives the world as it moves around. I wonder where crickets come from or cockroaches... something that I am in contact with constantly, everyday, and I tend to think about it because of that. Living, mysterious, and exciting. These are terms I most closely associate with in the natural world and how it appears. It's alive. It's mysterious and we don't understand it, and it's exciting. I tend to look at the natural world as being mysterious - that there is a lot that we don't know about it. It is exciting. I guess just because it is interesting to see how Nature works, and just how complicated it really is. I think that the more we know about it, the more enlightened we will be about ourselves and the world we live in, and the better chance we will have that we will be able to be more reasonable about our decisions that we make, and I think that kind of knowledge is hopeful and peaceful. And just from the basic scientific reasons, you never know what you're going to find when you go to study something, and just from past experiences, we've found that by looking closer, and with more detail, we discover more about how we, as humans, are constructed and how we relate to Nature and we're able to improve our quality of life because we are understanding Nature better. So, you can study it, I think, in many ways, and it is so amazing and interesting, to see and to experiment, that any curiosity that we have about it is a good enough reason to go and study it. There are aesthetic reasons. It is just pleasing to see how Nature works. I think that scientists are most involved in the study of Nature, as far as observing and trying to measure what is going on in the world, in turn to predict how things will change and what will happen. I think that is one of the functions that science really fulfills as far as a human enterprise. The studies themselves, I am optimistic about. How the studies are used, they are subject to all the human shortcomings and problems, but as far as doing the studies I am optimistic. I think that we have always got more to learn and that we can learn a lot from Nature.

Because of our dependence on Nature, just our existence, we need to study Nature, to learn more about it. We need to understand how things work in Nature because it is an important resource for us, to get our water, energy, food, and materials for making things from Nature. The resources that Nature contains is kind of unpredictable, because we don't really know what resources are there, that we can use. Because of that, we need to understand it as much as we can, so that we can protect it. It needs to be protected... and keep it so that it can maintain us and maintain civilization. Man has an impact on the natural world. Because the world is full of resources and powerful, man has also polluted and exploited it, even though it is powerful, and has taken the resources and used them for his own purposes and things. Although I think everybody has a sort of innate appreciation for life and for the natural world, when people have self-interest at stake, they tend to meet their own needs.

**About Mr. David:**

Mr. David is Anglo-American. He is a young, first year teacher. He had completed a biology degree followed by a secondary science teacher education program. He was enthusiastic about both the students and his subject. During the interview he made it quite clear that he had great concerns about pollution and the damage it does to Nature; and he strongly supported environmental efforts. He was enthusiastic about the contributions of science to environmental efforts. Mr. David also specifically replaced the word "God" in the interview with the word "beauty." He asserted, "I see *beauty* in Nature" and implied that he did not see God in Nature. He spent personal time in Nature hiking and camping.

# *Appendix C*
## Basic Interview Protocols

What follows are the basic Task One interview protocols as written for student interviews. The protocols for Task Two are basically the same. The protocols are appropriately modified for use with adults. An interview begins with a few introductory comments when the informant enters the room:

> "Hi,"
> "Have a seat,"
> "Thank you for coming," etc.
> "I think I have everything set up that I need except for your permission slip. Did you bring it with you?"
> "I want to assure you that this is not a test. We are going to have a conversation about your concepts of "nature" for a study that is being done at the university."
> "I'll be taping part of our conversation because I think we might be talking too fast for me to write everything down."

Subsequently, the interview proceeds as follows:

| INTERVIEWER DOES | INTERVIEWER SAYS |
|---|---|
| Spread out pictures and let student look at them. | "I've laid some pictures out in front of you, would you just take a moment to look at them?"<br><br>Pause |
| | "How would you define Nature or the Natural world?"<br><br>Pause |
| | "Is there any difference or are they essentially the same thing?" |
| If a distinction is being made - combine definitions or otherwise resolve. | |
| | "Would you say that all of these pictures depict nature or are there some that do and some that do not?" |
| Put up signs:<br>NATURE IS...<br>AND<br>NATURE IS NOT... | |
| | "We are going to go through a series of cards. I'm going to ask you to think about the words and then I'll ask you to comment about them." |
| Divide cards randomly into three groups. Use 1/3 at a time. Lay out 1/3 of the cards. | |

| | |
|---|---|
| | "Remember, what we are focusing on is what nature is. I want you to divide these cards into two groups. One group of words that you would use when talking about nature and one group that you would not use." |
| Repeat for each 1/3 of the cards. | |
| Take words in the "would use" pile and spread them out in front of interviewee. | |
| | "Some of these words may be about the same thing - do you want to lump any of these words together?" |
| Turn tape recorder on if it hasn't been turned on yet. | |
| Wait for student to pick. | "O.K., let's talk about these groups. Which of these would you pick out 1st to talk about "what nature is?"<br><br>"Fine" |
| Repeat chosen words for tape. | "What do these words have in common? Why have you put these words together?" etc. |
| Ask clarification questions if necessary. Avoid "Okay." | |
| | Why are you saying, "Nature is ___"? What do you mean when you say, "Nature is ___"? In what sense would you say, "Nature is ___"? What examples can you give me? |

| | |
|---|---|
| Ask non-directed questions that invited the interviewee to talk about why the terms were picked and what they mean. Ask for clarification and examples. Ask follow up questions were appropriate. | |
| Pull 1st group of words aside but keep them visible. | |
| | All right, we have this group aside now let's take a look at the rest. Which group would you pull out next? |
| | "Why did you pull <u>this</u> group out after this group?" |
| Go back through the question series as above. Look for conflicting words. As how first group relates to second or if they do. | |
| | Can you help me understand this? Why is it that on the one hand (name first or previous groups) and on the other its (name present group)? |
| Try for at least 4 groups of "<u>Nature is</u>." Do all if interviewee is prepared to continue - then go to "<u>Nature is not</u>." | |
| Repeat questioning as above. | |

# Index

## M



## N



## O



## P



## R



## S

## T



## V



## W

## *About the Author*

Dr. Cobern received his undergraduate degree in biology and chemistry from the University of California, San Diego (Revelle College) in 1971. After teaching high school science for several years, he studied for his doctorate in science education at the University of Colorado, Boulder receiving his Ph.D. in 1979. Dr. Cobern then spent a number of years as a faculty member at the Usmano Dan Fodio University of Sokoto, Nigeria. There he participated in the development of a new science teacher education program and the development of education programs for nomadic groups. His research was on the influence of traditional culture on learning in formal school settings. Dr. Cobern came away from that experience having given much thought to the notion that for many students science is a second cultural experience. From this basis in experience he developed a worldview theoretical framework for conceiving of science as an aspect of culture which warrants the assertion that meaningful science learning only occurs to the extent that scientific knowledge can find a cognitive niche within the everyday thinking of ordinary people. This theoretical work has been developed in several publications that have been studied and adopted by science education researchers in the Africa and Australasia who are struggling with cultural issues. In the USA, his work has drawn the attention of science educators working with Native Americans, African Americans, and religious communities.

Dr. Cobern has served on the editorial board for the *Journal of Research in Science Teaching*. He is the section editor for "Culture and Comparative Studies" for the journal *Science Education* and is a charter editorial board member for the *Canadian Journal of Science, Mathematics and Technology Education* He has guest lectured in Australia, Brazil, Israel, Japan, New Zealand, Nigeria, Norway, and South Africa. He is committed to the concept of culturally constructed, quality science education for all.

In 1996, Dr. Cobern joined the faculty at Western Michigan University where he holds joint appointments in the Department of Teaching, Learning & Leadership (College of Education) and the Department of Science Studies (College of Arts and Sciences).

Printed in the United States
87842LV00003B/344/A

9 780792 363446